Genetic Resources as Natural Information

T0256290

Demonstrating the shortcomings of current policy and legal approaches to access and benefit-sharing (ABS) in the Convention on Biological Diversity (CBD), this book recognizes that genetic resources are widely distributed across countries and that bilateral contracts undermine fairness and equity. The book offers a practical and feasible regulatory alternative to ensure the goal of fairness and equity is effectively and efficiently met.

Through a legal analysis that also incorporates historic, economic and sociological perspectives, the book argues that genetic resources are not tangible resources but information. It shows that the existing preference for bilateralism and contracts reflects resistance on the part of many of the stakeholders involved in the CBD process to recognize them as such. ABS issues respond very well to the economics of information, yet as the author explains, the application has been either sidelined or overlooked.

At a time when the Nagoya Protocol on ABS has renewed interest in feasible policy options, the author provides a constructive and provocative critique. The institutional, policy and regulatory frameworks constitute "bounded openness" under which fairness and equity emerge.

Manuel Ruiz Muller is Director of the Program of International Affairs and Biodiversity at the Peruvian Society for Environmental Law (SPDA), Lima, Peru. He is also Coordinator of the International Environmental Law course for the Environmental Law and Natural Resources Diploma at the Catholic University, Peru.

Genetic Resources as Natural Information

Implications for the Convention on Biological
Diversity and Nagoya Protocol

Manuel Ruiz Muller

Foreword by
Joseph Henry Vogel

Case Studies by
Klaus Angerer and
Omar Oduardo-Sierra

Routledge
Taylor & Francis Group

LONDON AND NEW YORK

from Routledge

First published 2015 by Routledge

2 Park Square, Milton Park, Abingdon, Oxfordshire OX14 4RN
711 Third Avenue, New York, NY 10017

Routledge is an imprint of the Taylor & Francis Group, an informa business

First issued in paperback 2017

British Library Cataloguing in Publication Data
A catalogue record for this book is available from the British Library

Library of Congress Cataloguing in Publication Data
Ruiz Muller, Manuel.
    Genetic resources as natural information: implications for the Convention on
    Biological Diversity and Nagoya protocol / Manuel Ruiz Muller; foreword by
    Joseph H. Vogel; case studies by Klaus Angerer and Omar Oduardo.
        pages   cm
    Includes bibliographical references and index.
    1. Germplasm resources–Economic aspects.   2. Germplasm resources–
    Government policy.   3. Convention on Biological Diversity
    (1992 June 5)   I. Title.
    S494.3.R85 2015
    333.95'34–dc23   2015012715

ISBN: 978-1-138-80194-3 (hbk)
ISBN: 978-0-8153-7895-2 (pbk)

Typeset in Bembo
by Out of House Publishing

To my loving family, Rosie, Manolo Jr., Alonso,
and small twins Lia and Teo, you are all my inspiration
and what keeps me going in life.

Contents

Illustrations

About the author and contributors

Manuel Ruiz is a lawyer working in environmental, natural resources and development law. He has specialized in biodiversity and intellectual property law and has had an active participation in national and international policy processes related to access to genetic resources and benefit-sharing (ABS) since 1991. In 1998, Ruiz was awarded a Darwin Fellowship to undertake research and collaborate with the Royal Botanic Gardens, Kew, on issues regarding its collections, repatriation and ABS. He has written extensively on genetic resources, traditional knowledge and misappropriation. In recognition of his work, the Peruvian Society for Environmental Law (SPDA) was awarded the Danielle Mitterrand Foundation Prize for 2014. Ruiz has been a consultant for many international institutions including FAO, IUCN, UNDP, UNEP and WIPO. His current work focuses on promoting a better reflection of the intricacies and complexities of science and technology as applied to genetic resources in law and policy, and the problems and challenges faced by existing ABS approaches both national and internationally. He is also a lecturer at the Pontificia Universidad Catolica del Peru where he teaches International Environmental Law. mruiz@spda.org.pe

Joseph Henry Vogel is Professor of Economics at the University of Puerto Rico-Río Piedras and has been a speaker at over 250 venues worldwide. He has served on the Ecuadorian delegation to the Conference of the Parties to the Convention on Biological Diversity (CBD) and to the Framework Convention on Climate Change (FCCC). Since launching *Genes for Sale* (Oxford 1994), Vogel has published extensively on the economics of appropriation, be it genetic resources, the atmospheric sink or even movie locations. He has participated in projects funded by the UNDP, USAID, IADB, World Bank and various NGOs. Abstract reasoning over appropriation has also resulted in innovation. On June 5, 2012, Vogel was awarded US Patent #8195571 for a "Web-Based System And Method For Preventing Unauthorized Access To Copyrighted Academic Texts", which was recognized as significant of the expansionary policy of the USPTO toward what is patent eligible. josephvogel@usa.net

Klaus Angerer is a lecturer in the Institute for the History of Medicine at the Justus-Liebig-University Gießen in Germany. His ongoing PhD dissertation examines the transformation of collected biological materials used for drug discovery and is based on fieldwork in academic and industrial laboratories as well as botanical gardens. Among his publications are "Frog tales – On poison dart frogs, epibatidine, and the sharing of biodiversity", "There is a frog in South America whose venom is a cure – poison alkaloids and drug discovery" and "Doing things with frogs – von der Erforschung von Froschgiften zu einer chemischen Ökologie". His research also addresses the likely consequences of regulations of the access to genetic resources and the fair and equitable sharing of the benefits arising from their utilization under the United Nations Convention on Biological Diversity (CBD) and the Nagoya Protocol. Klaus. Angerer@histor.med.uni-giessen.de

Omar Oduardo-Sierra graduated in Political Science from the University of Puerto Rico-Río Piedras where he is currently pursuing a Masters in Linguistics. His area of interest is cognitive analysis of discourse by Parties to the framework conventions of the United Nations. He is lead author of "Monitoring and Tracking the Economics of Information in the Convention on Biological Diversity: Studied Ignorance (2002–2011)" and co-author of "Human Pathogens as Capstone Application of the Economics of Information to Convention on Biological Diversity" and "La financiación y fungibilidad del Museo de Bioprospección, la Propiedad Intelectual y el Dominio Público" (The Finances and Fungibility of the Museum of Bioprospecting, Intellectual Property and the Public Domain). One of his research endeavors is a film documentary about the culture of coffee growers in Puerto Rico, an ever dwindling number. omaroduardo@gmail.com

Foreword

"On the Silver Jubilee of 'Intellectual Property and Information Markets: Preliminaries to a New Conservation Policy'"

Joseph Henry Vogel

Imagining alternatives is innately human: What if I had married my childhood sweetheart? What if I had chosen a different career? What if I had migrated? What if I had stayed? Such musings evoke a range of emotions and the possibilities are endless. A few remind us of our vulnerability. "What if I had been at that very spot a minute sooner?"

Although the above questions are self-centered, the template is not ego-bound. "What if" radiates out and can be scaled up. On the lecture circuit, paleontologist Stephen Jay Gould would imagine the boundary of the Cretaceous and Tertiary Periods, 65 million years ago. "What if the asteroid struck a few seconds later?" Traveling at a velocity of 20 km/second, the Gulf of Mexico would have absorbed the kinetic energy rather than the coastline of the Yucatán. The Age of Reptiles might not have ended, nor the Age of Mammals begun. Imagining the alternative is instructive. In the Age of The Bomb, something slightly different may trigger nuclear war and end in nuclear winter. Disarmament makes sense.

Far less dramatic than an asteroid or The Bomb are the synergistic reasons for mass extinction in the Age of Man: Habitat destruction, Invasive species, Pollution, Population and Over-harvesting rendering the acronym HIPPO (Wilson 2002: p. 50). The 1992 United Nations Convention on Biological Diversity (CBD) grappled with the variables H, I and O of HIPPO but left the P of Pollution for another convention and the P of population for another time. Three broad objectives were established in Article 1 of the CBD: "the conservation of biological diversity, the sustainable use of its components and the fair and equitable sharing of the benefits arising out of the utilization of genetic resources." Because the third objective enables the first two, the order was reversed *de facto* in the Conference of the Parties (COP). Yet, despite twenty years and twelve COPs, "the fair and equitable sharing of the benefits" remains elusive. Imagining a different trajectory of the CBD is the subject of this foreword; imagining the alternative outcomes, the subject of the Case Studies. The policy implications of a slightly different language – genetic resources as natural information – constitutes the body of the book.

Before musing over what if natural information had been the object of utilization in Article 1, one must understand how a foundational mistake – genetic resources as "material" – entered Article 2. Françoise Burhenne-Guilmin reflected on the events that culminated in the CBD in "Introduction to The Guide to the Convention on Biological Diversity," co-authored with Susan Casey-Lefkowitz (1994). Her first-hand observations are worthy of quoting at length:

> The UNEP Secretariat, assisted by a small group of legal experts, then prepared a first draft of the convention based on all the "elements" that had been produced so far. The formal negotiating process started in February 1991, when the group was renamed the Intergovernmental Negotiating Committee for a Convention on Biological Diversity (INC).
>
> The main issues were divided between two working groups ... Working Group II dealt with issues of access to genetic resources and relevant technologies, technology transfer, technical assistance, financial mechanisms and international cooperation. Progress was slow and negotiation difficult, especially during the final negotiating sessions. As time passed, the self-imposed deadline for signature of the Convention – the UNCED Conference in June 1992 – was approaching with alarming speed.
>
> The negotiations were often close to breaking down. Even on 22 May, the final day of the final negotiating session in Nairobi, it was not clear until the last moment whether the Convention would be adopted. Had the UNCED deadline not been present it is unlikely that a convention would have been adopted on that date. Yet in spite of this fact, and in spite of the tensions in negotiation, the number of signatures to the Convention in Rio on 5 June was unprecedented. The entry into force of the Convention, only 18 months after it was adopted was equally stunning.
>
> (Glowka *et al.* 1994: pp. 2–3)

The narrative reads like a cliff-hanger and returns us to my template. What if the negotiations had broken down? The imagination fires. Five sessions of the INC were held from February 1991 to May 1992. What if the UNCED deadline had not been present? Burhenne-Guilmin and Casey-Lefkowitz leave no doubt about what would not have happened: The Nairobi Final Act of the Conference for the Adoption of the Agreed Text of the Convention on Biological Diversity on May 22, 1992. Désirée Marielle McGraw draws the same conclusion in "The Story of the Biodiversity Convention: Origins, Characteristics and Implications for Implementation." She relates: "[T]he momentum created by a multiplicity of meetings, the completion of the climate change negotiations and the pending and highly public Rio Earth Summit served as incentives for concluding a biodiversity convention" (McGraw 2000: p. 17).

I have one degree of separation from the heady events that transpired in Nairobi in the spring of 1992. In February of that year, I attended the 4th World Congress on National Parks and Protected Areas held in Caracas, Venezuela. Cyril de Klemm chaired a session on the progress of the CBD or, better said, the lack thereof. A more suitable chair could not have been found. Ten years earlier de Klemm had presented "Protecting Wild Genetic Resources for the Future: The Need for a World Treaty" at the 3rd World Congress. In the intervening decade, his life-long vision had taken shape and he assumed center stage. Sporting a slightly crooked tie, the wispy 75-year-old seemed straight out of central casting. The role: the right-eous legal scholar who hails from the Continent.

Attendance at the session was low, embarrassingly so. Reminiscent of the teacher who chastises the students present for those absent, de Klemm bemoaned the lack of interest by the attendees to the 4th World Congress. Meanwhile the session across the hall was packed to overflow. Its topic? How to build footpaths and hand-rails in the bush. The mise-en-scène is engraved in my memory and dredges up a strange mix of metaphors. Only years later would I realize that the text emerging out of Nairobi could never have been anything more than jerry-built. Time did not permit it. The Spring of 1992 must have been a roller coaster of emotions for de Klemm. As McGraw relates: "Going into the final meeting on 22 May 1992, delegates had agreed on less than half of the Draft Convention: 27 out of 42 articles contained square brackets" (McGraw 2000: p. 15).

WAS FAUST LURKING?

What if the Convention had not been adopted at the Earth Summit, Rio '92? The cynic in me wants to agree with Melinda Chandler, the legal advisor to the US negotiating team and convenient villain in the narrative:

> It is regrettable that a legal instrument as ambitious as the Biodiversity Convention should suffer from basic conceptual and drafting deficiencies. The structure of the negotiations, the haphazard way in which crucial issues were considered, and the pressures of time contributed to a legal instrument which should cause distress for international lawyers and policy-makers.
>
> (Chandler 1993: p. 174)

But the optimist in me cannot agree. A reason existed for hope and still exists: the CBD is a framework treaty which evolves through the decisions of the COPs. Alas, twenty years have now lapsed and all the issues surround-ing "fair and equitable sharing of the benefits arising out of the utilization of genetic resources" remain contentious despite the Nagoya Protocol on the Fair and Equitable Sharing of Benefits Arising from the Utilization of Genetic Resources (Kamau et al. 2010). Even worse, new issues have been injected through the Protocol (West 2012), which also "suffers from basic conceptual and drafting deficiencies" precisely for the same reason: participants have to show something for all the time, effort and money expended. Faust re-surfaced and so, I oscillate. Perhaps the villain in Nairobi was a hero after all when she argued that "we as governments, lawyers, and policymakers can do better – much better – in crafting legal instruments that will advance environmental conservation" (Chandler 1993: p. 175).

Sometime in the early years of the CBD, "access to genetic resources" and the "fair and equitable sharing of the benefits arising out of the utilization of genetic resources" were conjoined and simplified. The acronym "ABS" emerged. It stands for a phrase that appears nowhere in the CBD: "access and benefit-sharing." To the ear of an economist, ABS sounds a lot like selling and buying, especially when the adjectives "fair and equitable" no longer modify "the sharing of benefits." So, why the euphemism? Why not an expression more amenable to "mankind in the ordinary business of life"? (Marshall 1890: p. 1). Why not "buy" or "sell"? Such directness leads to the heart of the matter: Who holds title?

The observations of Burhenne-Guilmin and Casey-Lefkowitz (1994) are again worthy of close examination:

> Since the early 1980s, several countries restricted access to the genetic resources under their jurisdiction, and the calls of developing nations for national controls over genetic resources have become increasingly louder. During the negotiation of the Convention on Biological Diversity, this point prevailed. As a result, article 15 recognizes that the authority to determine access to genetic resources rests with the national governments and is subject to national legislation.
>
> This evolution is based on the view that there is no *legal* reason to exempt genetic resources from the principle of national sovereignty over natural resources. But it is also grounded in practical reasoning: control over access to genetic resources gives the providing Party the opportunity to negotiate the mutually agreed terms for fair and equitable sharing of benefits required by article 15(7).
>
> (Italics added)

A legal reason to exempt genetic resources might have occurred to the negotiators had they first pondered the economic reason. When sovereignty is understood narrowly as bilateralism, the (mis)interpretation enables competition in genetic resources as long as the attribute of interest is diffused over jurisdictions. The asteroid whizzes and the former sentence requires unpacking. Users are interested in the "natural information" teased out of the biological sample through research and development (R&D) and will obtain the input from the cheapest Provider. To what extent Nature obliges, with many Providers, is an empirical question. Some bits of natural information are found in all life forms, e.g. ATP synthase, while others are ephemeral to a few individuals at a moment of time (see Case Study 1). Although one does not know the diffusion *a priori*, natural product chemistry documents enough redundancy to expect competition.

A cautionary note: ever since Adam Smith, the enlightened State accepts competition as a good thing – the invisible hand – bringing material bliss. That thinking is generally correct. The outstanding exception is information. Without *de jure* protection of information, the creator of (artificial) information or steward of (natural) information lacks incentives to create or to steward. So, limited-in-time intellectual property rights make sense for artificial information as do *sui generis* rights for natural information.

Did anyone express such thinking during the negotiations of the convention in Nairobi?

All degrees of separation now vanish. I had the good fortune to put the question to Burhenne-Guilmin at COP 6, held in the Slovak Republic in 1998. The Convention Hall was perched on a hilltop outside Bratislava and the building looked like a bunker. We were waiting for the elevator when I introduced myself and posed the question. Burhenne-Guilmin responded with mild astonishment. Didn't I know? They had indeed tried hard to establish a General Fund but the developing countries would have none of it! She half smiled, looked up and off to the side. She shrugged her shoulders in a gesture that is probably a human universal. The answer left me nonplussed. Although I do not doubt the accuracy of her impressions, I have nevertheless obsessed over the response. What she said meant that the negotiators from the developing countries acted against their own interests. That age-old advice "don't be your own worst enemy" rings in my head. Like the "what if's" about the asteroid, the homily radiates out and can be scaled up. The Global South had acted against themselves, decisively. Didn't anyone see the disaster coming?

Anniversaries are a time of reflection. Jeff McNeely, former director of the IUCN, said on the occasion of the 10th Anniversary of the CBD, which also coincided with the twentieth anniversary of the 3rd World Congress of the IUCN:

> Cyril de Klemm called for a convention on genetic resources. He saw this as a means of securing free and open access, while also charging for international trade in such resources with the income earned going into an international fund that would support conservation activities in the developing world. The IUCN Environmental Law Centre continued to develop articles for inclusion in such a convention and when Mostafa Tolba convened a small group in his office in 1988, we were ready with some reasonably well fleshed-out ideas. Once negotiations began in earnest, some of our ideas were overtaken by other considerations.
>
> (CBD News Special Edition 2002: p. 5)

In other words, the small group converged on the implications of the economics of information without recognizing genes as information. What if they had recognized genes as information? Would they still have let the Global Fund be "overtaken by other considerations?"

McNeely's quote invites revision. The openness of "bounded openness" can be interpreted as the "free" in de Klemm's call for "free and open access"; the bounds, as "charging for international trade." My memory of the bunker returns; the elevator door opens. I cannot remember the exact words said but I do recall that Burhenne-Guilmin implied that the opposition from the developing world was due to the "Global" in "Global Fund". We get on the elevator and my memory fades. What if de Klemm had advocated a convention on natural information? The small group which met with Tolba were trained to think abstractly; surely, the *economic* reason would have resonated and been communicated in Nairobi. The Global Fund could have become isomorphic with "bounded openness", thereby meeting the dual criteria for any successful ABS scheme that the legal scholar Charles R. McManis would spell out years later: "theoretically sound foundations and [the capability] of relatively low cost implementation" (McManis 2004: p. 427).

I have long thought: what a pity! Reductionism – genetic resources as natural information – would have triumphed had it only been heard in Nairobi. It would have been a small step to suggest a distribution of the royalties based on the geography of the natural information utilized in any intellectual property. Twenty years and as many refereed publications later, I realize just how wrong I was. Logic and evidence do not necessarily persuade, especially as any small group expands to include those who are not persuaded by the power of logic and evidence. Only recently do I perceive how a "tragedy of unpersuasive power" (Vogel 2013) penetrates all of the letters of HIPPO and becomes more destructive than any one of them. The tragedy lends itself to a derivative mnemonic which I will call THIPPO.

No metaphor is exact and the asteroid is no exception. Asteroids make sudden impact, attested by the well-defined boundary of the Cretaceous and Tertiary. In contrast, THIPPO gnaws relentlessly in human time. But time is relative. E.O. Wilson imagines its passing at speeds which range from the fastest (biochemical) to the slowest (geological) (Wilson 1984: p. 42). To rescue my asteroid, I must deploy another metaphor, also from Wilson: the movie projector. Let us see the twelve COPs in geological time. We ramp up the projector. Twenty years are compressed into what Wilson would describe as "less than an eye blink in the starry message of the cosmos" (Wilson 2014: p. 54). The asteroid is once again an apt metaphor: its trajectory tracks THIPPO but full impact is still several seconds away. When did the asteroid start to veer off course? When did people start thinking along the lines of bounded openness rather than bilateral negotiations over genetic material?

The answers are not easy to ascertain. Because many ways exist to express the same idea, one must search a variety of expressions to determine the date of its debut. To complicate matters, "bounded openness" is not a single idea but a set of interrelated ideas. What is the critical mass to affect the solution of ABS? When were the ideas published? And what really constitutes publication?

The question of the date of debut is central to aligning incentives and thereby achieving conservation and sustainable use, the first two objectives of the CBD. Showing a lack of due diligence by either the negotiators in Nairobi or the delegates to the COPs will help future delegates rethink the object of utilization in Article 1 and correct the foundational flaw of Article 2. The earlier the debut, the weaker is any invocation of *stare decisis* (stand by the decision) for bilateralism. Did the critical mass for "bounded openness" appear before the presentation of the CBD for signature at the Earth Summit, Rio '92? Before COP 1 in 1994? Before COP 2 in 1995? Before COP 12 in 2014? The asteroid can only veer with the elimination of the "T" of THIPPO.

But I am getting ahead of myself. The elements for "bounded openness" can be ordered into sequential steps. The first ten constitute the critical mass sufficient for the solution and the order approximates their relative importance:

1. Recognition of the utilization of genetic resources as the utilization of natural information, which invites the application of the economics of information and the justification of rents;
2. Incentives through the extension of property rights over natural information in a multilateral system;
3. Disclosure of utilization in the transmittal of applications for intellectual property;
4. Establishment of a Global Fund to hold royalties in escrow;
5. Imposition of a royalty rate with revenues destined to the Global Fund;
6. Recognition of redundancy of natural information at different taxa as an empirical question;
7. Recognition of the determination of the diffusion of natural information across taxa as a transaction cost subject to change, decaying with technological advances;
8. Recognition of the determination of the geographic distribution of the information dispersed across taxa as a transaction cost subject to change, also decaying with technological advances;
9. Dispersal of royalties to the countries of origin, proportional to the relative holdings of the natural information, when the costs of determinations (7) and (8) are inferior to the sum in escrow for the natural information utilized;
10. Dispersal of the sum collected in the Global Fund to the infrastructure required to make the determinations whenever the costs of so doing are superior to the sum collected at the moment the intellectual property right expires on the utilization.

Beyond the ten essential elements lie another five which facilitate the acceptance and efficiency of "bounded openness". Unlike the previous ten elements, the order of the additional five does not indicate relative importance:

11. Determination of the geographic diffusion of the natural information among landowners in a country of origin with dispersal whenever the sum is superior to the cost of the determination of geographic share;
12. Recognition of public domain for all natural information already commercialized when the system begins;
13. A design of penalties for non-disclosure of the use of natural information to align incentives;
14. Recognition that the solution does not generate a market value to be integrated into any quixotic calculation of Total Value of Biodiversity but instead creates a Galbraithian "countervailing power" against HIPPO;

15. A negotiation of royalty rates between User and Provider countries based on a matrix of relevant characteristics of utilization.

To determine the date of the debut of any element, one must also establish what constitutes publication. E.O. Wilson provides a clue: "Science grows in a manner not well appreciated by non-scientists: it is guided as much by peer approval as by the truth of its claims" (Wilson 2012: p. 276). Not surprisingly, the Harvard professor emeritus sets a high bar for scientific growth through publication. Let's define "The Platinum Standard" as publications which are peer reviewed by experts in the field. A notch below is "The Gold Standard" constituting scholarship from think tanks in the form of bulletins, newsletters and discussion papers. Continuing downward, newspapers and magazines would constitute "The Silver Standard". At the bottom is Straw, which are the instantaneous uploadings on the web.

A Google search of the words "benefit", "sharing" and "genetic resources" yields, as of the date of this writing, 529,000 hits. No one can be accused of lack of due diligence for not sifting through all that straw. However, if we consider only those publications which are peer reviewed – a contraction by about three orders of magnitude – we hazard misattribution of ideas which may have earlier taken root by standards more accessible to the stakeholders of the COPs. To complicate matters, testing a publication will generate results shy of the critical mass, no matter what standard we choose. For example, the core idea of "incentives" (Element 2) can be inferred in *Biophilia*, when Wilson writes "The only way to make a conservation ethic work is to ground it in ultimately selfish reasoning ... An essential component of this formula is the principle that people will conserve land and species fiercely if they foresee a material gain for themselves, their kin, and their tribe" (1984: pp. 131–132). However, the intrigued reader will not so easily infer any of the other nine elements from *Biophilia*. The learned reader will also balk at the originality of "ultimately selfish reasoning" and think of Adam Smith's *The Wealth of Nations* (2007 [1776]) which is also "not a wholly original book" (Heilbroner 1979: p. 49). Indeed, ascertaining the originality of any one idea could conceivably take us back to the Greek Philosophers. Nevertheless, ascertaining the originality of the critical mass – all ten elements together – is within our lens of resolution.

Coming closer to "bounded openness" and the presentation of the CBD at Rio '92 is "Property Rights for Plants" by Roger A. Sedjo, an economist with the think tank Resources for the Future. The four-page article appears in the in-house news journal *Resources*. One immediately infers Elements 2 and 11 in the following excerpt:

Under a system in which the concept of property rights was extended to include species not now known or utilized, newly discovered natural genetic resources would become the property of the political state in which the resource resides. In principle, the state would be free to declare all such resources as the property of the state, or it could grant private property rights to individuals or to corporations that discover the genetic resources. Having

ownership of the resources, the owners – public or private – could be expected
to have an interest in their long-term preservation and development.

(Sedjo 1989: pp. 2–3)

Albeit suggestive of the critical mass for bounded openness and at the cusp of the
negotiations in Nairobi, Sedjo missed the economic meaning of genetic resources
as information and even inveighs against its overarching implication – a biodiversity cartel:

> The country where the resource resides could negotiate an exclusive agree-
> ment with a firm, or allow a number of firms to utilize the resource under a set
> of mutually agreed-upon conditions. Agreement might be reached as part of
> bilateral negotiations or as the result of a competitive bidding process. Should
> a particular germplasm be discovered in several countries simultaneously, the
> potential users would be free to negotiate the best deal possible with the coun-
> try of their choice … [T]he longer a monopolist withholds the germplasm
> from the market, the greater is the possibility that events will compromise the
> favorable initial bargaining situation. Where several countries have the same
> unique germplasm resource, the possibility for collusion and the formation of
> a cartel exists. However, cartels have been historically unstable, and the possi-
> bilities for finding alternative germplasm resources are likely to be substantial.
>
> (Sedjo 1989: p. 3)

The excerpt is remarkable inasmuch as Sedjo had written, just the previous year,
about the value of species as "a repository of genetic information that someday
may have direct commercial and/or social value." The quote appeared in the highly
visible anthology *Seeds and Sovereignty*, three years before Nairobi (Kloppenburg
1988: p. 296). Although Sedjo referred to genetic resources as information, he
would proceed to treat them as if they were material. A pull-out quote in the
Resources article projects the equivocation in large type: "Genetic material markets
could function just as markets do for other resources" (Sedjo 1989: p. 3). The observation is true but problematic; the price in a competitive market will settle at the
marginal cost of collecting samples, virtually nothing.

Speculation can be instructive. What if Sedjo had reflected on genes as information? I believe he could have deduced the critical mass of bounded openness for
his chapter in *Seeds and Sovereignty*. Had that happened, I would not be writing this
foreword now. The asteroid would have veered, sharply. On what basis do I speculate?
Critics of intellectual property regimes have long disparaged patents with the epithet
"monopoly". Sedjo was undoubtedly aware of the defense for time-limited monopolies as a means to recoup the fixed costs of R&D. It would have been low-hanging
fruit to justify some sort of "oligopoly property right" to conserve "repositories of
genetic information" and explore the institutional exigencies. I speculate that Sedjo
missed the obvious for a reason that is Skinnerian. Economists are reflexive with
any reference to "oligopoly", "cartel" or "rent-seeking behavior". The conditioning

begins in the introductory course, more specifically, Chapter 9 in the classic textbook *Economics* (Samuelson and Nordhaus 2005). I count myself fortunate not to have been encumbered by an undergraduate education in economics; I studied chemistry.

In 1990, I published "Intellectual Property and Information Markets: Preliminaries to a New Conservation Policy" in the newsletter for the Centre for International Research on Communication and Information Technologies (CIRCIT), located in Melbourne, Australia. I was affiliated with CIRCIT as a summer research fellow. On the Silver Jubilee of "Intellectual Property" in 2015, a Google-Scholar search reveals not one single citation. But I am not disheartened. Rereading the piece, I am happily surprised at the prescience of the title. It is indeed preliminary and falls short of the critical mass needed to solve ABS. Nevertheless, "Intellectual Property" demonstrates more elements than any other publication as of that date (Elements 1, 2, 6–8 and 11). In just one page, I advocated "an extension of property rights to include genetic information" and concluded that

> [h]abitats do not correspond to landownership patterns and therefore, genetic information, is likely to be owned jointly among landowners. The contractual relationships between these landowners and the industrial users of genetic information must be carefully considered. My work at CIRCIT will elaborate on these issues and outline incentive structures for landowners, corporate boards and systematists.
>
> (Vogel 1990)

The following summer, I published "The Intellectual Property of Natural and Artificial Information" (Vogel 1991). Its major point was that intellectual property and genetic resources are homologous in information. Re-reading the article, I see that I had added Element 4 to the argument made the previous summer:

> Protection of natural information requires unique institutions to deal with joint ownership. Whereas artificial information is the product of one inventor or one group of inventors, natural information is diffused over land owned by many individuals. The creation of a property right over natural information would be the creation of a right shared in common by all owners of the habitat. To establish claims to royalties, an international biological inventory would be required.
>
> (Vogel 1991: p. 7)

The two articles were still too sketchy to be of any real use for the negotiators in Nairobi, even if miraculously they were somehow to have been read. Missing were four elements, 3, 5, 9 and 10, necessary for the solution. Nevertheless, the co-directors of CIRCT, Don Lamberton and Bill Melody, must have sensed that I was on to something. They supported my efforts to flesh out the full argument. After the following summer fellowship in 1992, I stayed on at CIRCIT and worked feverishly on the manuscript, some 170 pages. At the AIC Conference on

Biodiversity held in Sydney on November 16–17, 1992, I launched the special limited edition *Privatisation as a Conservation Policy.* The softcover book was designed to look "in-house" and not foreclose the possibility of publishing the same text with an academic publisher. Oxford University Press of New York accepted the manuscript in 1993 and it appeared the following year as *Genes for Sale.*

And the four elements missing from the newsletter articles? Were they present in *Privatisation?*

They were present with differing degrees of elaboration. For example, Element 3 appears *en passant* when I write that "corporations reveal the genetic information used in their products to the inventory" (Vogel 1992: p. 39) whereas Elements 5 and 9 are somewhat detailed:

> The delineation of a commons for genetic information is complex inasmuch as landowners may have valuable information in organisms on their parcel of land but not know who else has this same piece of information … [T]hree types of transaction costs would be incurred: (1) the identification of the taxon at which the genetic information is distributed (2), the identification of other landowners who have that same piece of genetic information on their land and (3) the design and implementation of a scheme to exclude non-paying users from access to that information. Clearly, these transaction costs are enormous … The only way to lower these costs is to capture economies of scale and reduce the average cost per landowner of establishing his share of the genetic information commons. Theoretically, this can be done by following some simple steps to reduce the aforementioned transaction costs (1)-(3). Although the sequence of steps may be simple, each step is a complex task: (1) the identification of the taxon at which the genetic information is distributed; this will require molecular analysis of the organism for which the GCF [genetically coded function] has been commercialized and then molecular analyses of organisms from the same race, species, genus, etc., to measure the distribution of that GCF across taxa, (2) the identification of the landowners who have that same piece of information on their land; this will require not only a database but also a biological inventory with records for each land title and (3) the design and implementation of a scheme to exclude non-paying users from that information; this will require that industries which use natural genetic information in new products identify that usage, declare its dollar value, and remit a royalty to the commoners.
>
> (Vogel 1992: p. 56)

Element 10 was broached in Chapter 9 entitled "Who will finance the Gargantuan Database?" (Vogel 1992: p. 95). Because the costs of the system could be greater than the monies collected, I argued that the monies should, in such instances, "be used to diminish the fixed costs of the Gargantuan Database" (Vogel 1992: p. 96). Element 11 also has its own chapter, Chapter 6: "Genesteaders." In contrast, Element 12 only appears in a Footnote, number 8 of Chapter 5: "Like an expired patent or copyright, genetic information that is already commercialised for a specific function would be in the public domain" (Vogel 1992: p. 38). Years later the last three additional elements for bounded openness would enter the research stream. Element 13 can be found in "Reflecting Financial and Other Incentives of the TMOIFGR: The Biodiversity Cartel" (Vogel 2007a) and Element 14, in "White Paper: The Successful Use of Economic Instruments to Foster the Sustainable Use of Biodiversity" (Vogel 1997). Element 15 is the most recent addition and was first discernible during the Online Discussion Group on Article 10 of the Nagoya Protocol, from April 8 to May 24, 2013, through the ABS Clearing House Mechanism.

Before the formal negotiations began in Nairobi in 1991, three other academics were converging on bounded openness, although under different nomenclatures. Whether the negotiators in Nairobi exercised due diligence turns on the publication date of the critical mass. Of the other three publications, the earliest is Timothy Swanson's Discussion Paper "The Economics of the Biodiversity Convention" (Swanson 1992).

Unlike Sedjo, Swanson recognized the implications of genetic resources as information. The Discussion Paper satisfies Element 1 when he writes: "The presence of variation is information; uniformity is the absence of information. Therefore, the diversity inherent in biological resources contains information simply by definition" (Swanson 1992: p. 13). Element 2 is also easy to find as Swanson uses the same word, "incentive". In Element 4, he deploys a synonym, "International Fund" for "Global Fund". However, the presence of Elements 5 and 6 requires inference, and Elements 7–9, an even greater capaciousness in interpretation. Element 10 would be a stretch, as it requires that the reference to "World Heritage Convention" (Swanson 1992: p. 27) also includes possible modalities of the financial mechanism of the "World Heritage Fund", not cited. Totally absent is "the disclosure of utilization in the transmittal of applications for intellectual property", which is ranked third in the ten elements of the critical mass. Without it, "bounded openness" cannot work. In summary, applying the filter of Elements 1–10 to Swanson's discussion paper, we find an absence of Element 3, extreme difficulty in discerning Element 10 and ambiguity in discerning of Elements 7–9. So, a convergence on the critical mass did not happen despite its publication in the same year as *Privatisation*. However, Swanson *et al.* added many of the missing elements soon thereafter, in a background study paper for the FAO, "The Appropriation of the Benefits of Plant Genetic Resources for Agriculture" (Swanson *et al.* 1994).

Two other works stand out which suggest the broad outlines of "bounded openness": Christopher Stone's 42-page article, "What to Do about Biodiversity: Property Rights, Public Goods, and the Earth's Biological Riches" (Stone 1995) and Barbara Laine Kagedan's 176-page report "The Biodiversity Convention, Intellectual Property Rights, and the Ownership of Genetic Resources: International Developments", prepared for the Intellectual Property Policy Directorate Industry Canada (Kagedan 1996). The former provides enough overarching statements to infer the elements of the solution, and the latter, enough institutional detail to substantiate them.

As Swanson, Stone, Kagedan and I were imagining institutional arrangements for ABS, other economists were taking a different path. The divergence is reminiscent of the early history of economic thought: "For Malthus the issue was the immensely important one of 'How Much Is There?' For Ricardo it was the even more explosive question of 'Who Gets What?'" (Heilbroner 1979: p. 99). Bruce A. Aylward (1993), R. David Simpson *et al.* (1996), Gordon C. Rausser and Arthur A Small (2000), and many others, have all grappled with "How Much Is There?", in other words, "What are genetic resources worth for R&D?". Their very different estimates reflect slightly different assumptions and again make the metaphor of the asteroid apt. In contrast, Swanson, Stone, Kadegan and I were asking the more explosive "Who Gets What?", I believe that I have also gone further and answered "How Will They Get It?". The reason for my follow-up question is simple: if the transaction costs in deciding "Who Gets What?" are greater than the answer to "How Much Is There?" then "What is the Point?".

Nonetheless, the model builders were right to insist on some estimate of value. Ironically, an indicator was established about the same time the first model made its debut. The polymerase chain reaction (PCR) revolutionized biotechnology and its discoverers won the 1993 Nobel Prize in Chemistry. By 2005, the expired patent on PCR had earned US$2 billion (Fore *et al.* 2006). That fact provokes reflection: Just one piece of natural information, an enzyme, from one species, *Thermus aquaticus*, could have generated US$300 million for countries of origin at the royalty rate I suggested in 1992, a startling 15 percent. The sum is three times the six-year budget of International Barcode of Life (Genome Canada 2011: p. 8), which could help achieve Elements 6–8 needed for "bounded openness". With 100 million or more species on the planet, the question "How Much Is There?" need not be answered with any precision.

The publication of the necessary and sufficient elements for "bounded openness" – all ten elements – occurred with the launch of *Privatisation* on November 17, 1992, five months after the submission of the final draft of the CBD (May 22, 1992). So, we cannot accuse the negotiators in Nairobi of a lack of due diligence. We cannot reach the same conclusion, however, regarding the delegations to the twelve COPs. Before COP 1 met in 1994, and ever since, various scholars have independently published the solution for ABS in increasingly fine detail through venues which meet the Gold and Platinum Standards. The legal concept of *laches* may be appropriate; it never set in. Sometime in the first decade of the COP discussions, the lack of due diligence morphed into studied ignorance (Oduardo-Sierra *et al.* 2012). From about COP 9 onward, one will hear stakeholders begrudgingly acknowledge the logic and evidence for "bounded openness" but, taking a deep breath, also dismiss it as a solution which has come too late. Somehow they say this with a straight face.

The book before us by Manuel Ruiz appears twenty-five years after the first elements of "bounded openness" surfaced in "Intellectual Property and Information

Markets: Preliminaries to a New Conservation Policy." The asteroid has scorched
Planet Earth badly but need not continue its course to full impact. The Case Studies
are thought experiments which imagine the ABS that might have been, and still can
be. Ruiz opens the book with a Turkish proverb that is both timely and timeless.
As long as biodiversity is threatened with extinction through the misalignment of
incentives, we must turn back.

Preface

I first came across "access to genetic resources" and "the fair and equitable sharing of benefits" (ABS) in 1990 while still a law student and uncertain about my future. I was hooked immediately. I had the good fortune to be an intern at the Peruvian Society for Environmental Law (SPDA), which by mere chance also began to participate in the national, regional and international ABS discussions. Twenty-five years later, I continue to work at SPDA, now as the Director of the International Affairs and Biodiversity Program, and still on ABS, albeit not exclusively.

A lot seems to have passed since then. But maybe this is not the case. The Convention on Biological Diversity (CBD) was approved; dozens of ABS policies and laws are in place; literally hundreds of guidelines, books and papers related to ABS have been written; two international ABS agreements (the FAO International Treaty and the Nagoya Protocol) have also been adopted; well documented and researched ABS case studies have been produced; courses and training modules imparted; millions of dollars have been invested in ABS cooperation programs, projects and meetings, to enumerate a few occurrences.

During the early 1990s, I was introduced to some economic and legal works on genetic resources and ABS by Professors Joseph Henry Vogel, Tim Swanson and Christopher Stone, but quite frankly disregarded them initially because of their complexity (so I thought). Also because I was in some way already immersed in and riding on certain conceptual and regulatory trends advocating for contracts in ABS, particularly through work in the Andean Community in 1994. At the same time, I became interested in the science and technology applied to genetic resources while a Darwin Fellow at the Royal Botanic Gardens, Kew, and started to strongly sympathize with some of the concerns and plights scientists faced in understanding and complying with new and evolving control-based ABS policy, legal and institutional frameworks.

Convinced of the need for and usefulness of contracts to regulate ABS, I relatively rapidly came to realize that the dream for equity and justice in the use of genetic resources would not come to fruition through bilateral, one-on-one negotiations, for reasons I will come to explain in this book. And so I somehow migrated to a new line of thought.

This book is, above all, about a personal and modest intellectual journey through ABS and the acceptance of early errors which I have had no problem in admitting publicly. I firmly believe ABS needs to move in another direction to deliver what it is meant to deliver: conservation, equity and fairness.

I do not intend to take any credit for the ideas and suggestions proposed in this book. I have simply tried to synthesize and present them in a way which I hope is better received in ABS quarters. My contribution is marginal at best.

I met Joseph (Joe) Vogel personally in the mid-1990s and started studying more carefully some of his and a few other colleagues' economic ideas and proposals in ABS. I always wondered why the ideas of Stone, Swanson and Joe were not taken more seriously earlier. Maybe it's been a matter of personalities, circumstances, who knows? What I can say is that I have not seen (as yet) a more coherent and grounded approach to ABS – but which certainly moves in another direction from current ABS trends. I am still to see a convincing technical rebuttal to some of these proposals.

In tandem with my own building doubts and concerns regarding the path taken by ABS as a whole for over a decade, I was able to – I would like to think – acknowledge the need for an alternative policy framework which is more practical, feasible, technically grounded and economically justified. Having said this, I have to also admit that this "new" approach to ABS will be (and already is) contested politically in light of its implications on national sovereignty and countries of origin. Furthermore, the ABS approach proposed in the book means the admission that progress in ABS is almost a mirage. Useful in some respects, but with no practical or substantial content in terms of equity and fairness. Much less for the conservation of biodiversity.

A note of caution: the book does not address traditional knowledge (TK) of indigenous peoples and communities. Though I am fully conscious of the importance of TK and its relevance in the context of ABS, it is my modest view that this is an issue which would require separate research and another methodological approach. Some references to TK are made and Case Study 2 offers an initial suggestion on how TK may be protected or integrated into the bounded openness proposal outlined throughout the book.

Finally, my goal with this book is not to be politically correct, or incorrect for that matter, but rather, be honest with my beliefs and try to work through a logical pattern which is the result of careful reflections, talks, interviews and reading over the past two decades. The tangible result is what you hold in your hands (something I truly appreciate) and the reader can judge for him/herself whether these ideas and reflections are worth the read.

Manuel Ruiz
December 28, 2014
Lima, Peru

Acknowledgments

Though I am the author of and fully responsible for the book, I truly feel it is a collective effort. I'd like to thank first and foremost my friend and ABS colleague Joseph Henry Vogel for his support during the writing of this book. I am also deeply grateful to Omar Oduardo-Sierra and Klaus Angerer for their insightful comments and substantive inputs to the text, especially with two illustrative case studies. At different moments during writing I sought guidance and comments from colleagues and long-time friends. My thanks extend to Graham Dutfield, Seizo Sumida, Allan Jimenez, Carla Bengoa, Jorge Caillaux, Pierre du Plessis, Carlos Correa and Manuela Gonzalez.

I would like to acknowledge the Peruvian Society for Environmental Law (SPDA), my home institution and where I have been able to learn and freely think about ABS (and other issues). Its continued support for work started in 1992 has been invaluable for my professional growth and understanding of ABS dilemmas and the world in general. My deep appreciation to my colleagues and friends there.

My sincere appreciation also extends to Federico Burone from the International Development Research Centre (IDRC) and Tim Hardwick from Routledge for their continued support of this book. Finally, over the years, many organizations have been supportive of my work and have given me the confidence to continue exploring and finding my way. My special thanks to IDRC, The MacArthur Foundation and GIZ for their support.

Acronyms and abbreviations

ABS	Access and benefit-sharing
CBD	Convention on Biological Diversity
CGEN	Council for the Genetic Patrimony
CIRCIT	Centre for International Research on Communication and Information Technologies
CITES	Convention on the International Trade in Endangered Species
COP	Conference of the Parties
DNA	Deoxyribonucleic acid
EPO	European Patent Office
ETC	Group Erosion, Technology and Concentration
EU	European Union
FAO	United Nations Food and Agriculture Organization
FDA	US Food and Drug Administration
FOEN	Federal Office of the Environment and Nature
FTA	Free Trade Agreement
GATT	General Agreement on Tariffs and Trade
GBIF	Global Biodiversity Information Facility
GEF	Global Environment Facility
GLMMC	Group of Like-Minded Megadiverse Countries
GMBSM	Global multilateral benefit-sharing mechanism
GMO	Genetically modified organism
IARCs	International Agricultural Research Centers
iBOL	International Barcode of Life Initiative
INBIO	National Biodiversity Institute
IP	Intellectual property
ITPGRFA	International Treaty on Plant Genetic Resources for Food and Agriculture
IUCN	World Conservation Union
JPO	Japanese Patent Office
MAT	Mutually agreed terms
MTA	Material Transfer Agreement
NGO	Non-governmental organization

NIH	National Institutes of Health
NMR	Nuclear magnetic resonance
OPEC	Organization of the Petroleum Exporting Countries
PBR	Plant breeders' rights
PCR	Polymerase chain reaction
PGRFA	Plant genetic resources for food and agriculture
PIC	Prior informed consent
R&D	Research and development
RNA	Ribonucleic acid
SBSTTA	Subsidiary Body on Scientific, Technical and Technological Advice
SMTA	Standard Material Transfer Agreement
SPDA	Peruvian Society for Environmental Law
TEV	Total economic value
TK	Traditional knowledge
TRIPS	Agreement on Trade Related Aspects of Intellectual Property Rights
UNCED	United Nations Conference on Environment and Development
UNCTAD	United Nations Conference on Trade and Development
UNEP	United Nations Environment Programme
USPTO	United States Patents and Trademarks Office
WIPO	World Intellectual Property Organization
WTO	World Trade Organiation
WWF	World Wide Fund for Nature

Introduction

Of all the issues addressed by the Convention on Biological Diversity (CBD), the most intractable is access to genetic resources and the fair and equitable sharing of benefits (ABS). Since the entry into force of the CBD in 1993, ABS has never subsided as the bone of contention among Parties. Traditional knowledge (TK) has also persisted as a contentious issue over the years but will not be addressed in depth in this book, for reasons explained in the Preface.

Both time and resources have been heavily invested and, according to some critics, not necessarily wisely so in the development of legal frameworks, analysis of ABS dimensions in large- and small-scale projects in bioprospecting, frequent international, regional and national meetings and an array of capacity building courses. A torrent of publications now exists that deal specifically with ABS, the CBD, the International Treaty on Plant Genetic Resources for Food and Agriculture (ITPGRFA 2004) and the Nagoya Protocol on Access to Genetic Resources and the Fair and Equitable Sharing of Benefits Arising from their Utilization (2010).[1] Why then another book?

Equity and fairness in ABS have become elusive. So why have they not been achieved, especially in the case of fairness and equity in the sharing of monetary benefits? This is a fundamental question, is answerable, and merits exploration.

The broad concept of ABS and its implications can be traced to the early 1970s in the discussions surrounding the Green Revolution, plant genetic resources for food and agriculture (PGRFA), and the emergence of International Agricultural Research Centers (IARC). The Keystone Dialogues, the works by Jack Kloppenburg and the Crucible Group, followed soon after and were essential to catalyze and further debates.

Ownership and increased used of technologies over genetic resources were becoming more controversial, especially from a policy perspective. Nevertheless, the CBD is a watershed. Access and benefit-sharing, the third of only three inter-linked objectives of the CBD,[2] placed the issues of control and rights over genetic resources center stage. Coincidentally, biotechnology would take off simultaneously with the patent over the polymerase chain reaction (PCR) by Cetus Corporation in 1983, which derives from an aquatic microbial genetic resource collected in Yellowstone National Park.[3] Everyone could make the connection.[4] Genetic resources were of immense value and under imminent threat.

When the CBD was signed at the Earth Summit (Rio '92), countries classified as "developing" perceived an opportunity to become full participants in the benefits to be generated by transnational biotechnology, located primarily in the developed countries. Discrete events can be identified since the signature of the Convention that have become critical to the trajectory of ABS in the twenty-first century and beyond. They are summarized in Table I.1.

Above all, developing countries perceived ABS as a strategic maneuver long over-due. They now had a forum, the CBD, to broaden the scope of debates from PGRFA within the United Nations Food and Agriculture Organization (FAO) context to any genetic resource in their jurisdiction. With the much vaunted sovereignty over genetic resources, the meaning of "biopiracy" was almost self-evident: a misappropriation of genetic resources and related traditional knowledge.[5] The neologism fired indignation.

History was easily retrievable. The contact of Europeans with the peoples of the Middle and Far East and Africa in the fourteenth century and the Americas in the fifteenth century also meant contact with genetic resources of the lands "dis-covered" and the beginning of globalization. The Vavilov centers of domestication became particularly important in terms of their contribution to unrestricted flows usually from the New to the Old World.[6] With the expansion of biotechnology in the 1990s, developing countries realized that the value of genetic resources went beyond agriculture and hinged on monopoly intellectual property of their utiliza-tion, largely in biotechnological patents and plant breeders' rights (PBR).

Almost as soon as the CBD entered into force in November 1993, developing countries responded with a control-based ABS policy and regulatory framework. The Philippines and the Andean Community translated ABS principles into spe-cific instruments founded on the CBD notions of "sovereignty", "benefit-sharing", "prior informed consent" (PIC) and "mutually agreed terms" (MAT). In hindsight, the urgency to develop ABS frameworks came at the expense of scant reflection about the essence of genetic resources and the potential consequences of error in misclassifying that nature.

As of 2014, at least 60 countries[7] and regional blocs had either an ABS legal frame-work in place or were at some initial or advanced stage of developing one.[8] Many look strikingly similar in their substantive content, approach and formal structures, including those from Brazil, Costa Rica, India, Malaysia, Panama, South Africa, the Philippines, the Seychelles, the Andean Community (including specific regulations in Bolivia, Colombia and Peru), the African Union and the European Union (EU).

Experience to date shows that countries with ABS frameworks in place face considerable challenges in implementing them and achieving a measurable fair and equitable participation in benefits derived from the access to and use of their gen-etic resources. Repeated failure by diverse countries is a good indicator that some-thing is intrinsically wrong with the ABS approach in general, and that fundamental change is needed.[9] The central theme of this book is that correction is not only possible but easily within reach.

In other words, the problem with existing ABS frameworks should not be attrib-uted to complex administrative procedures or limited institutional capacities of

Table 1.1 ABS–biotechnology–IP–TK: events and effects

Year	Event	Effect
1980	US Supreme Court Decision Diamond v. Chakrabarty	Possibility of patenting biotechnological products and processes – derived from genetic resources – broadened
1980	Bayh-Dole Act (amendments to intellectual property (IP) legislation in the US)	Favors private investment in research and development in biotechnology innovation generated by public and private research centers and universities – public research becomes increasingly "enclosed"
1992	Adoption of the CBD	Linkages between genetic resources, biotechnology, IP and traditional knowledge (TK) placed on a broader international agenda/platform (specific references to each issue in CBD text) – PIC and MAT affirmed
1993	US signs the CBD but qualifies its interpretation (it has not ratified the CBD to date)	Strong resistance by industrial sectors in the US, especially the biotechnology and pharmaceutical sectors which prefer non-restricted access, to ratification – CBD ABS and other obligations arguably[10] not applicable to the US
1994	End of the Uruguay Round of the General Agreement on Tariffs and Trade (GATT)	Intellectual property included at the last minute into the GATT agreements through the Agreement on Trade Related Aspects of Intellectual Property Rights (TRIPS) – minimum international standards for patents over biotechnology defined as well as sanctions for non-compliance defined as part of the World Trade Organization (WTO) rules
1996	First ABS laws enacted (Andean Community and the Philippines)	ABS contracts and bilateralism (PIC/MAT) become part of regulatory frameworks and begin to influence other processes worldwide
1999	Initial assessment by the World Intellectual Property Organization (WIPO), of the need for an international TK protection regime	Traditional knowledge and genetic resources become part of the international IP "global issues" agenda of WIPO – synergies developed with WTO, CBD and FAO forums
2001	Adoption of the FAO International Treaty on Plant Genetic Resources for Food and Agriculture (ITPGRFA)	A standard Material Transfer Agreement (MTA), a global fund and a multilateral system for benefit-sharing is established for a defined set of PGRFA
2010	The Nagoya Protocol on Access to Genetic Resources and Benefit Sharing is signed	Affirmation of sovereignty and bilateralism through ABS contracts, with an opening for a global multilateral mechanism/system for ABS (Articles 10 and 11) in the case of transboundary, shared resources, etc.
2014	Entry into force of the Nagoya Protocol	The EU considers a regulation to support the Nagoya Protocol implementation from the perspective of the User

public competent authorities per se. These may exist but are not causal. The central problem lies in what we consider essential conceptual flaws which are rooted in the CBD and which have spilled over to national and international ABS legislation over time. They are three: (1) an erroneous definition in the CBD of "genetic resources", (2) excessive emphasis on sovereignty, and (3) bilateralism (expressed in ABS contracts, PIC and MAT), and its misplaced concern over physical access to genetic resources. These have contributed toward ineffective and inefficient approaches to ABS. Overarching the three conceptual flaws is intransigence despite the framework nature of the CBD.

The first flaw is easiest to grasp. CBD misdefined "genetic resources" as "material" and the mistake resonates in ABS policy and regulatory frameworks. Only the intangible or informational dimension of genetic resources is of interest to biotechnology (Vogel 1991; Swanson 1992; Stone 1995; Kagedan 1996; Pastor and Ruiz 2008). By ignoring or sidelining the essential informational element of genetic resources, policymakers and regulators developed frameworks which amplify the foundational flaw and thereby sidelined the relevant and applicable field of economics, namely, the economics of information. In other words, any acknowledgment of information as the object of access would have guided development of a policy and regulatory approach for ABS that would have been radically different in form and content, both internationally and nationally.

Time has not been kind to the misclassification of genetic resources as "material". Scientists routinely highlight the critical importance of information-based tools in analyzing genetic resources and whole disciplines have emerged rooted in information (e.g. genomics, synthetic biology, proteomics, meta-genomics, bioinformatics, and so on).[11] The capacity of the Parties to the CBD to celebrate path-breaking research on genetic resources as information while ignoring any policy based on genetic resources as information has not gone unnoticed and invited accusations of "doublethink".[12] Accusations of double-speak easily follow as failure on ABS is passed off as success. The lack of discussion of the obvious and simple solution is sidelined because it seems obscure and complicated.

So, the book departs from the status quo in the implementation of the ABS discourse, and tracks a largely ignored literature that begins two years before the signature of the CBD at Rio '92. Departure risks alienation of stakeholders who have sometimes committed decades in painstaking negotiation, for very little in light of the foundational flaws. One colleague who responded to the idea of an alternative approach to ABS based on cartelization and bounded openness expressed skepticism, "unless there is a complete shift in thinking". Such a shift is not as ambitious as it may seem. Once the definitional flaw of the CBD is recognized, the shift is afoot and the obvious and simple solution can no longer be dismissed as somehow obscure and complicated.

To explore the depth and breadth of the shift, the book is divided into six chapters (including a chapter on conclusions and recommendations) and two case studies which include two thought experiments relevant to ABS – in the context of natural information. These chapters are structured in a logical order to allow the reader to understand stepwise the rationale. The obviousness and simplicity of the economic argument lends itself to compression. Professor Joseph Henry Vogel, one of the advocates for change, has compressed the essential elements of the economic argument into the Foreword, which celebrates the silver jubilee of its beginning with a publication dated 1990. Having literally and figuratively accompanied the journey of the COP, his reflections on what went wrong suggest a larger problem that he calls "the tragedy of unpersuasive power" where the CBD is "exemplary".

Having participated in ABS discussions and processes since 1991, I also am struck by the steadfast adherence to the current ABS model. However, the years and decades have worn down considerably the resistance to the critics and provide the openings for change. How much change? This will depend on the will of ABS actors to recognize the commission of error and embrace correction and embark on a new process.

Having once been an advocate of bilateral contracts, I have learned first-hand the meaning of "sunk costs", which is well expressed in the Turkish proverb "no matter how long you have traveled on the wrong road, turn back." Perhaps my affiliation with a think tank in the developing world, the Peruvian Society for Environmental Law (SPDA), affords me the flexibility to reflect, revisit and modify some of my own initial ideas and advocate for a new and viable ABS regime which goes against and defies the status quo.

Chapter 1 analyzes "natural information" (the term launched by Vogel in 1991) in lieu of "genetic resources" as the object of access. The substitution is the bifurcation point for a shorter path to benefit-sharing. The chapter reviews how scientists have long understood genes, molecules, metabolites, proteins and compounds as information, now in stark relief with the "-omics revolution", genomics, proteomics, metabolomics, glycomics, as well as bioinformatics and related disciplines.[13] The recognition of genes as information can also be found in the early 1990s in the refereed publications of some legal scholars who contemplated biodiversity, genes and ABS, such as James Boyle and Christopher Stone, as well as economists who grappled with the question of value including David Simpson, Bruce Aylward and Gordon Rausser. Amazingly, the actual policy, economic and legal implications of genes as information have never been on the negotiating table at any of the twelve Conferences of the Parties (COP) of the CBD and only marginally through submissions to its Subsidiary Body on Scientific, Technical and Technological Advice (SBSTTA) gatherings.[14] Nevertheless, the implications have appeared in parallel sessions or side events but which garner limited attention. Few delegates bother attending these meetings.[15]

The overarching implication is so simple as to be reducible to one paragraph. One begins with exclusion, a criterion for private goods in the marketplace. Physically, one cannot "fence in" information in any cost-effective manner and, therefore, must rely on the legal instruments of intellectual property (e.g. patents, copyrights, trademarks). Natural information presents an additional challenge: most vehicles of its transmission, namely, organisms, belong to species and genera that are highly dispersed and distributed across jurisdictions.

Through competition among Provider[16] countries, the price of genetic resources is driven down to the cost of collection, conceivably the price of a few kilos of dried leaves or a few biological samples. Bilateral ABS contracts and regulatory frameworks under which they are negotiated translate into "paying peanuts for biodiversity" as Professor Peter Drahos so eloquently suggests (Drahos 2014: pp. 138–143).[17] Hence, the secrecy over contracts or Material Transfer Agreements (MTAs) is a *modus operandi* to diffuse ire. Only a global and multilateral institutional structure which sets the price can guarantee any significant monetary benefit to the countries of origin from which the natural information was extracted.

Many scientists highlight that minute differences exist at the molecular level, which may make a genetic resource – natural information – unique to a specific site and even moment in time. Counterintuitively, such cases do not support bilateral negotiation as is developed in the thought experiment of Case Study 1. When there is only one country of origin, the terms of benefits should still be set by a multilateral approach. One justification is avoidance of the transaction costs of negotiations. But the larger reason is to concentrate significant benefits on species which may be vulnerable to changes in land use if found in just one country. Thus the multilateral mechanism aligns incentives for *in situ* conservation in particular, in a fashion that bilateral negotiations and existing regulatory and institutional frameworks cannot. Time works against ephemeral natural information and MTAs are time intensive.

Chapter 2 focuses on regulatory and policy trends in ABS derived from the CBD. The common elements of the common approach run counter to the policy implications of genetic resources as natural information. Confidentiality clauses in bilateral contracts and regulatory frameworks restrict the debate about fairness and equity in the sharing of its benefits. Secrecy is at the expense of transparency and the dilemma emerges: which rights have supremacy? One would think that in light of fairness and equity, transparency would rule. But the COPs have lionized "mutually agreed terms", which is shorthand for "confidential business information". Should secrecy trump via "mutually agreed terms", evaluation of fairness and equity in the sharing of benefits becomes unknowable, whistleblowers notwithstanding.

Chapter 3 offers an overview of the CBD based on the critical issues which should be revised – either as a modification to the text of the Convention or in a Decision of the Parties or a concerted interpretation. The chapter begins with the identification of historic milestones in the COPs, including how sovereignty over genetic resources and intellectual property have interacted and often driven discussion among the Parties.

The chapter also contemplates the concept and misconceptions of "sovereignty" and its unintended consequences for developing countries. Bilateral and standardized contracts foment competition among Providers to the benefit of Users. Were the consequences of sovereignty also unforeseen by the Users? Any response in the affirmative would mean a gross miscalculation of the transaction costs. The relative paucity of ABS agreements means the Users have also been frustrated (Carrizosa *et al.* 2004; Robinson 2010). Taxonomy and basic research have been

especially affected since the initial era of ABS regulations, with delays in processing permits, unconcluded administrative procedures, excessive requirements, etc.
(Hoagland 1998; Grajal 1999; Mansur and Cavalcanti 1999).[18] The chapter also
addresses the issue of sovereignty in the light of the simple economics of information. Understanding this section is key to understanding the rationale behind the
proposal on "bounded openness" elaborated in Chapter 5.

Chapter 4 offers some of the possible explanations for the resistance of the
ABS community to take note and examine the ideas suggested in this book. They
include the familiar causes of path dependency, agent–principal problems due to
speculation about groupthink as instinctual human evolution as a eusocial species. Resistance also means that advancements in science and technology have also
not been assimilated. Despite the "-omics revolution" in molecular biology and
its application to taxonomy with the advent of the International Barcode of Life
(iBOL), one hears advocacy for basically the same approach as that launched at
Rio '92! The "Medicine Man" paradigm, epitomized by the 1992 movie starring
Sean Connery, is passé (to a great extent) and long gone or at least has given way to
more sophisticated forms of research and development (R&D).[19] Both Parties and
stakeholders of the bilateral approach are heavily invested in what has proven to be,
above all, anachronistic.

With time, cracks have emerged in the resistance, possibly because of the entry
of new faces and the departure of the old. A few members of the ABS community
express willingness to look at ABS from an alternative perspective. The Nagoya
Protocol (Article 10, Global Multilateral Benefit Sharing Mechanism – GMBSM –
and Article 11, Transboundary Cooperation) constitutes an opportunity for a vigorous discussion of the issues and ideas suggested in the book.[20] Article 10 calls – with
a fair share of qualifiers – for a wholly new international and multilateral ABS
regime in the case of resources shared by more than one country, despite tortured
interpretation to the contrary. This book attempts to rise to that challenge.

Chapter 5 argues that a framework of "bounded openness" may resolve almost all
the problems and challenges which ABS presents, both conceptually and practically.[21]
From the rather complex distinction between commercial and non-commercial
research, to *ex situ* collections and concerns of retroactivity, a (new) legal framework
based on bounded openness offers a robust solution to the concerns afflicting the
ABS community and Parties to the CBD. Albeit politically challenging, the solution offers the best hope to realize the third objective of the CBD and secure the
interests of countries in genetic resources, fair and equitable benefit-sharing and,
most importantly, conservation and sustainable use. A cost-effective technical solution is within reach and "logic should prevail" in the building of ABS institutional
and legal frameworks (Ruiz *et al.* 2010). Two decades and twelve COPs are enough.

Under this new approach, "openness" means that "genetic resources should flow
freely with notable exceptions (e.g. endangered, invasive and pathogenic species)"
while "boundedness" means that "a significant royalty would be levied on commercially successful intellectual property and shared among countries of origin pro-
portionally to habitat" (Vogel 2007b). Notice that the concept of "boundedness"

refers to an *ex post* situation and not to an *ex ante* regulatory constraint or procedure as the substantial elements.

In the end, the policy reduces to a question of money. To avoid competition among Providers, the royalty rate must be fixed. The original suggestion was 15 percent of the net sales of biotechnological products and services derived from natural information (Vogel 1994), 2 percent of which would be for genetic material (Vogel 2007b). But the variance in utilizations may justify a table of royalties based on a combination of characteristics in the value added (Vogel 2013). Case Study 2 explores monopoly intellectual property other than patents which now fly under the radar. The important point is not the royalty percentage ultimately negotiated through the COP for any type of intellectual property but its uniformity for any particular category of utilization. The insight emerged from a spirited exchange during the Expert Online Discussion Groups on ABS convened in 2013 by the Secretariat to the CBD.[22] It also finds convergence in the 2014 Guidelines for ABS of India[23] and similar proposals before the Brazilian legislature in 2014.[24]

An institution would need to be designed to support effective distribution of benefits to countries of origin – for conservation of habitats and ecosystems. Note that this is not merely a General Fund. The iBOL initiative, the Global Biodiversity Information Facility (GBIF), and similar initiatives, could play a critical role in determining the geographic dimensions of habitats and presence of species and hence, support the definition of the proportion of royalties to be shared among countries. Both Providers and Users should realize that bounded openness is also an enlightened approach; access will be vastly facilitated for *all* purposes, thereby increasing the probability of utilization and commercially successful "hits".

To complement the chapters as an integral part of the book a thought experiment is tendered: how would ABS have proceeded were the policy in place "bounded openness" rather than ABS contracts and bilateralism? The case studies were carefully chosen: *Epipedobates anthonyi* (poison dart frog) and *Lepidium meyenii* (Andean maca). The contrasts are striking and confirm the final views and suggestions in Chapter 6.

NOTES

1 The Nagoya Protocol entered into force on October 12, 2014.
2 Article 1 of the CBD (Objectives) establishes that its objectives "to be pursued in accordance with its relevant provisions, are the conservation of biological diversity, the sustainable use of its components and the *fair and equitable sharing of the benefits arising out of the utilization of genetic resources, including by appropriate access to genetic resources and by appropriate transfer of relevant technologies*, taking into account all rights over those resources and to technologies, and by appropriate funding" (italics added). Note that benefit-sharing is the nucleus of the provision and third objective, while access and technology transfer are instrumental toward the essential goal of equity and fairness.
3 For the story of *Thermus aquaticus* (Taq), see ten Kate, K., Touche, L. and Collins, A. *Yellowstone National Park and the Diversa Corporation. A Benefit Sharing Case Study.* Submission to the Executive Secretary of the Convention on Biological Diversity by

the Royal Botanic Gardens, Kew, April 22, 1998. Available at www.cbd.int/financial/bensharing/unitedstates-yellowstonediversa.pdf

4 Robin Pistorious recounts the history of the plant genetic resources movement, the Green Revolution and development of IARCs in Pistorious, R. (1997) *Scientists, Plants and Politics. A History of the Plant Genetic Resources Movement.* Rome: International Plant Genetic Resources Institute. Regine Andersen also undertakes a comprehensive review of the history of policies and institutional frameworks applicable to plant genetic resources in Andersen, R. (2008) *Governing Agrobiodiversity: Plant Genetics and Developing Countries.* Aldershot: Ashgate.

5 The neologism "biopiracy" was coined by Pat Mooney and promoted through his non-governmental organization (NGO) the Rural Advancement Foundation International (RAFI), renamed as Erosion, Technology and Concentration (ETC Group). Vandana Shiva embraced the term in her book of the same title. The political clout of the concept was instrumental in driving and influencing national and regional regulatory ABS processes, especially in the Andean Community, Brazil, Costa Rica and the Philippines. Nevertheless, no universally accepted definition exists. The Third and Final Complementary Disposition of the Peruvian Law 28216 (Law for the Prevention of Biopiracy, 2004) is perhaps the only legislation which captures the contested dimensions of the concept, defining "biopiracy" as:

> non authorized nor compensated access by third parties to biological resources and traditional knowledge of indigenous peoples without the appropriate authorization, in contravention of the Convention on Biological Diversity and existing norms. This appropriation may occur through physical control, through intellectual property rights over products which incorporate these illegally obtained elements, or in some cases through claims.
>
> (Author's translation)

The definition covers both physical access per se and intellectual property over related innovations, including those derived from traditional knowledge. Over time, the concept has been largely replaced by the less charged notions of "misappropriation" or "non-authorized access." For a comprehensive analysis of biopiracy and its implications, see Robinson, D.F. (2010) *Confronting Biopiracy: Challenges, Cases and International Debates.* London and Washington DC: Earthscan.

6 The conceptual framework developed by Vavilov in the early twentieth century helped explain the origin of crops and their diversification centers. Henry Hobhouse provides a comprehensive history of the diverse impacts of the global exchange of germplasm (genetic resources, seeds, whole plants). Although his review covers only six plants (cotton, coca, potato, quinine, sugarcane and tea), the analysis could be extended to many other plants useful for cosmetics, food, dyes and pharmaceuticals. These include maize, rice, soya, the rubber tree, the Pacific yew, the rosy periwinkle, the Neem tree and countless others. Animals, microorganisms and fish species have also been subject to global exchanges over time, with similar effects. See Hobhouse, H. (1999) *Seeds of Change. Six Plants that Transformed Mankind*, 4th edn. London: Papermac.

7 The CBD Secretariat database on ABS laws and regulations indicates there are 57 countries with ABS legislation in force as a specific law, regulation or measure and 7 regional blocs with ABS regimes in place. See www.cbd.int/abs/measures/

8 For an overview of these developments, see Cabrera-Medaglia, J., Perron-Welch, F. and Rukundo, O. (2012) *Overview of National and Regional Measure on Access to Genetic Resources and Benefit Sharing. Challenges and Opportunities in Implementing the Nagoya Protocol.* Centre for International and Sustainable Development Law (CISDL). 2nd Edition, July. Available at http://cisdl.org/biodiversity-biosafety/public/CISDL_Overview_of_ABS_Measures_2nd_Ed.pdf

9 "Failure" will be explored throughout the book in different contexts: (a) ABS policies and laws which can never lead to fairness and equity due to internal contradictions;

(b) bioprospecting projects and R&D processes which do not result in any commercial or industrial products; (c) bioprospecting projects and R&D which do result in commercially successful products but which are not linked to or regulated by ABS frameworks under which they should be governed; (d) projects and R&D which are interrupted or abandoned due to State inaction, administrative hurdles, and so on.

10 Though the official US government's position appears intransigent with regard to the CBD ratification, some US universities and botanical gardens as well as the National Institutes of Health (NIH), abide by ABS principles in their interactions with foreign universities and research institutions. The New York Botanical Gardens and Missouri Botanical Gardens have developed policies supportive of ABS principles and others have participated in projects for adapting microbial collections to CBD principles (e.g. MOSAIIC or the Pilot Project for Botanical Gardens on Common Policy Guidelines on ABS).

11 Twice Nobel Laureate in Chemistry Fred Sanger (1958 and 1980) is largely credited with having launched the science of genomics, which was further developed by Craig Venter, Francis Collins, and others. For a rapid review of the "-omics revolution", see Dalton, J. (2013) *Synthetic Biology and the "Omics" Revolution*. The Center for Issue & Crisis Management. United Kingdom. May 2013. Available at www.issue-crisis.com/uploads/Articles/SyntheticBiologyandOmics.pdf. For details on the history of bioinformatics, see Thampi, S. *Bioinformatics* (date not indicated). Available at http://arxiv.org/ftp/arxiv/papers/0911/0911.4230.pdf

12 Omar Oduardo-Sierra, Barbara Hocking and Joseph Henry Vogel quantify how many publications in the academic literature on ABS reference genetic resources as information yet ignore the economics of information. See Oduardo-Sierra, O. *et al.* (2012) "Monitoring and Tracking the Economics of Information in the Convention on Biological Diversity: Studied Ignorance 2002–2011." *Journal of Politics and Law* 5(12): 29–39.

13 Following Sanger, the works by Watson and Crick, Venter and Collins and others have perfected and opened the pathways to new forms of undertaking science, research and development, applying powerful computational tools in the light of evolutionary theory.

14 Paul Oldham and the International Civil Society Working Group on Synthetic Biology have submitted documents to COP and SBSTTA which touch on the informational dimension of genetic resources, but stop short of identifying its implications for ABS policy. See, for example, Oldham, P. (2004) "Global Status and Trends in Intellectual Property Claims: Genomics, Proteomics and Biotechnology." Submission to the Executive Secretary of the Convention on Biological Diversity. Center for Economic and Social aspects of Genomics. United Kingdom (available at www.cesagen.lancs.ac.uk/resources/docs/genomics-final.doc) and The International Civil Society Working Group on Synthetic Biology. *A Submission to the Convention on Biological Diversity's Subsidiary Body on Scientific, Technical and technological Advice (SBSTTA) on the Potential Impacts of Synthetic Biology on the Conservation of Biodiversity*. October 17, 2011 (available at www.cbd.int/doc/emerging-issues/Int-Civil-Soc-WG-Synthetic-Biology-2011-013-en.pdf).

15 On November 3, 1996 during the IUCN Global Biodiversity Forum at COP 3 (Buenos Aires), Vogel presented "The Rationale for a Cartel over Biological Diversity in Bioprospecting." A month later he presented the commissioned White Paper "The Successful Use of Economic Instruments to Foster the Sustainable Use of Biological Diversity: Six Cases from Latin American and the Caribbean" for the Summit of the Americas on Sustainable Development, Santa Cruz de la Sierra, Bolivia, December 6–8, 1996. Case number VI was entitled "The Impossibility of a Successful Case in Bioprospecting without a Cartel" (Vogel 1997). Since Vogel's first publication on the topic in 1990, he counts over 200 venues where he has spoken on the application of the economics of information to genetic resources as natural information.

16 The concepts of "Provider" and "User" throughout the text are used broadly. The Provider is usually a country. User is almost always an institution, or a researcher, or a company, yet sometimes in common usage, may also be a country.

17 Australia is a very illustrative example: highly industrial or post-industrial User as well as megadiverse Provider with strong governance that should be able to negotiate significant royalties for access to and use of its unique natural information. The following table discloses royalties as an increasing function of revenues captured:

Monetary benefits in an ABS agreement between the Government of Australia and an undisclosed User

Purpose of the product	Gross exploitation revenue received in one calendar year (US$)	Threshold payment (% of gross exploitation revenue)
Pharmaceutical, nutraceutical or agricultural	<440,000	0
	440,000 to 4,400,000	2.5
	>4,400,000	5.0
Research	>175,000	2.5
	or	0
	<88,000	1.0
	88,000 to 2,600,000	3.0
	>2,600,000	
Industrial, chemical, diagnostic or other	>175,000	1.5
	or	0
	<88,000	1.0
	88,000 to 2,600,000	2.0
	>2,600,000	

Note: This table is adapted from WIPO. *Draft Intellectual Property Guidelines for Access to Genetic Resources and Equitable Sharing of the Benefits Arising from their Utilization.* Consultation Draft, February 2013, p. 21, and Australian dollars have been converted to US$.

18 Early in the discussions, many scientists foresaw the chilling effects ABS frameworks could have on field research in the Andean Community and Brazil. In both Peru and Brazil, illegal collecting of specimens was criminalized leading to accusations of paranoia. The Philippines also experienced similar concerns from its scientific community upon implementing Executive Order 247 in the mid-1990s. The scientific community routinely complains about the burdens that current ABS approaches place on research. For the plight of field researchers in Brazil, see Mansur, A. and Cavalcanti K. (1999) "Xenofobia na Selva: Paranoia Envolvendo Biopirataria Prejudica Pesquisas Científicas com Especies Brasileiras." *Revista Veja*, Ed. 1611, Año 32, No. 32–33, August, pp. 114–118. For a general view about the impacts of the Andean Community legislation on ABS on research, see Grajal, A. (1999) "Régimen de Acceso a los Recursos Genéticos Impone Limitaciones a la Investigación en Biodiversidad en los Países Andinos." *Interciencia* 24(1): 63–69. For a reflection on the effects of the CBD and ABS on systematics collections worldwide see Hoagland, E. (1998) "Access to Specimens and Genetic Resources: An Association of Systematics Collections Position Paper." ASCOLL, Washington DC. (ASCOLL is now part of the Natural Science Collections Alliance.)

19 *The Medicine Man* was a popular 1992 movie starring Sean Connery and Lorraine Braco, portraying the life of an ethnobotanist in the Amazon rainforest who had discovered the cure for cancer from a medicinal plant used by indigenous peoples, but subsequently lost which species was the source. Meanwhile the agricultural frontier was advancing, bulldozing the forest. Although the mise-en-scène was the Brazilian Amazon, the film

was actually shot in Veracruz, Mexico. In some frames, one can even glimpse the slopes of the Orizaba volcano. Scenery is also a form of natural information and its falsification in the visual arts has been termed "geopiracy". See Vogel, J.H., Robles, J., Gomides, C. and Muñiz, C. (2008) "Geopiracy as an Emerging Issue in Intellectual Property Rights: The Rationale for Leadership by Small States." *Tulane Environmental Law Journal* 21 (Summer): 391–406.

20 See initial reflections on this in Tvedt, M.W. (2011) *A Report from the First Reflection Meeting on the Global Multilateral Benefit Sharing Mechanism.* FNI Report 10/2011. Fridjt Nansen Institute, ABS Capacity Development Initiative. Available at www.fni.no/doc& pdf/FNI-R1011.pdf

21 Only one scenario is imaginable where bounded openness may *not* offer a satisfactory solution: value is added to natural information by an inventor who places the product in the public domain. If the Provider country does not use the open access good, it may object to the loss of the opportunity to receive a benefit by another User who would have sought IP protection. When this unlikely scenario is weighed against all other beneficial scenarios, it is insignificant.

22 In order to conduct a broad consultation on Article 10 of the Nagoya Protocol (as a result of the mandate in COP Decision XI/1B), the Secretariat of the CBD organized an Online Discussion Group through the Access and Benefit-Sharing (ABS) Clearing-House from April 8 to May 24, 2013. The Online Discussion focused on a set of questions (contained in Annex I to Decision XI/1). The synthesis of the report prepared by the CBD Secretariat is available at www.cbd.int/doc/?meeting=ABSEM-A10-01

23 India has adopted a similar sectoral approach in its Guidelines for Access and Benefit Sharing of August 2014. See Mazoomdaar, J. (2014) "Centre Sits on Royalty Slabs for Bio Resources, Loses Rs 25,00 cr a Year." *NATION*, November 19. Available at http://indianexpress.com/article/india/india-others/centre-sits-on-royalty-slabs-for-bio-resources-loses-rs-25000-cr-a-year/

24 See PL 7735/14 "Projeto da biodiversidade vai a comissão geral com várias polêmicas em aberto" (November 6, 2014), available at www2.camara.leg.br/camaranoticias/noticias/POLITICA/477144-PROJETO-DA-BIODIVERSIDADE-VAI-A-COMISSAO-GERAL-COM-VARIAS-POLEMICAS-EM-ABERTO.html

Chapter 1

The relevant nature of genetic resources

"Normal science" becomes "paradigm shift"

The language of the CBD allows interpretation for the implementation of ABS principles. Its capaciousness reflects the compromises that were necessary in the drafting of the final text in Nairobi in 1992 under the auspices of UNEP. Curiously, no capaciousness was allowed *de facto* for the definition of "genetic resources", despite the treaty being "framework" in nature.[1] Delegates and almost all academics have treated genetic resources as they are defined in the treaty, i.e. essentially as "things".[2]

Article 2 of the CBD defines "genetic resources" as "genetic material of actual or potential value." The definition includes known values for a specific genetic resource as well as those yet to be realized, thereby in practice including all genetic resources. In turn, "genetic material" is defined as "any material of plant, animal or microbial or other origin containing functional units of heredity." Genetic material is thus understood as "material from any biological source where units of heredity are operating or having a function" (Schei and Tvedt 2010) (Box 1.1).

The language of Article 2 seems cut and pasted from that found for "plant genetic resources" in the earlier FAO International Undertaking for Plant Genetic Resources (1983) which defines "plant genetic resources" as the reproductive or vegetative propagating *material* of plants (Article 2.1). The ITPGRFA went on to define "plant genetic resources for food and agriculture" as "any genetic material of plant origin of actual or potential value for food and agriculture" and "genetic material" as "any material of plant origin, including reproductive and vegetative propagating material, containing functional units of heredity." Although copying is usually extremely efficient, one should make sure that what is copied is indeed correct. Repetition of a mistake is also a mistake (see Table 1.1).

Time has a way of revealing mistakes, unintentionally. Research may identify new functions of genetic resources and streamline the development of technology, new goods and services in fields previously unimaginable. The past two decades are a good example of superlative progress in the life sciences and molecular biology in particular. The CBD definition stresses the physical dimension of genetic resources accessible from *in situ* and *ex situ* sources. The Nagoya Protocol and various national

Table 1.1 The cascading of a category mistake

	Definitions	Focus
The FAO International Undertaking (1983)	*Plant genetic resources*: "the reproductive or vegetative propagating material of plants"	Reproductive and vegetative propagating material (tangibles – seeds, stems, etc.)
The CBD (1992)	*Biotechnology*: "any technological application that uses biological systems, living organisms, or derivatives thereof, to make or modify products or processes for specific use"	Biological systems (group of organs – material), living organisms (material), derivatives (material)
		Biotechnology (in its use) applies to the informational element of these systems, organisms and derivatives
	Genetic material: "any material of plant, animal, microbial or other origin containing functional units of heredity"	Materials (tangible)
	Genetic resources: "genetic material of actual or potential value"	Genetic materials (tangible)
The ITPGRFA (2001)	*Genetic material*: "any material of plant origin, including reproductive and vegetative propagating material, containing functional units of heredity"	Genetic materials (tangible) – containing DNA
	Plant genetic resources for food and agriculture: "any genetic material of plant origin of actual or potential value for food and agriculture"	Genetic materials (tangible)
The Nagoya Protocol (2010)	*Biotechnology*: "any technological application that uses biological systems, living organisms, or derivatives thereof, to make or modify products or processes for specific use"	Biological systems (group of organs – material), living organisms (material), derivatives (material)
		Biotechnology (in its use) applies to the informational element of these systems, organisms and derivatives
	Derivative: "a naturally occurring biochemical compound resulting from the genetic expression or metabolism of biological or genetic resources, even if it [the derivative] does not contain functional units of heredity"	Biochemical compounds (material)

Table 1.1 (cont.)

Definitions	Focus
Utilization of genetic resources: "to conduct research and development on the genetic and/or biochemical composition of genetic resources, including through the application of biotechnology as defined in Article 2 of the Convention [the CBD]"	Research (e.g. through biotechnology) applies to the informational element of these systems, organisms and derivatives

ABS laws and regulations also focus on access to the tangible dimension, literally codifying the definitional mistake worldwide.[3]

Tangibles under these definitions include biochemical compounds and derivatives even though the intangible dimension of the resource is where, in terms of biotechnological developments, the value lies. In the case of the Nagoya Protocol a "cascade interpretation" involving access, biotechnology and utilization of genetic resources is now deployed to broaden the scope of ABS to include derivatives or derived products (as they are often called) – but still in their material and tangible form.[4] Throughout this book I will suggest that whether we refer to genetic resources, a derivative, or a derived product (e.g. a resin, a crude oil or a natural extract), the focus should be on the natural information which is stored in the material vehicle or tangible support.[5]

In light of this shortcoming in the CBD and the Nagoya Protocol, and reminiscent of the original concerns of Stone, Swanson and Vogel expressed in the 1990s, some scholars are revisiting the CBD definition of "genetic resources". Schei and Tvedt rightly conclude that:

> If the concept of genetic resources is understood only narrowly, in senses related to the original or current state of knowledge, the ABS system [sovereignty, PIC, MAT] may not be able to capture the future potential value of genetic material [natural information], not least when it is used in or as a basis for synthetic biology or other new bio-economic technologies [including modern biotechnology and all of the "omics"]. An International ABS Regime could maintain a broad and dynamic understanding of the concept of "genetic resources".
>
> (Schei and Tvedt 2010: p. 8)

Dynamism in understanding the concept of "genetic resources" seems a roundabout way to correcting the category mistake of the CBD. For the purpose of sharing benefits fairly and equitably, the object of access could be identified as "natural information". Hence, "genetic resources" should either be redefined as "natural information" or more elegantly "the object of access" identified as natural information.

The Nagoya Protocol extends the scope of ABS to research and development of genetic and/or biochemical compounds. As a result, any natural substance extracted from a biological source – provided research is undertaken on their biochemical or genetic composition through biotechnology – may be subject to ABS rules, depending on the specific national legislation.[6] The willingness to extend the definition engenders hope of a willingness to correct the category mistake.

Indeed, experts are increasingly paying attention to and taking note of the informational dimension of genetic resources from which the potential value of any biotechnology arises. It is natural information plus human creativeness through biotechnology and the "omics" disciplines which drive R&D.

In 1994, the Board of Appeals of the European Patent Office (EPO) explicitly indicated that DNA is "a chemical substance which carries *genetic information* and can be used as an intermediate in the production of proteins which may be medically useful" (italics added).[7]

In the context of ABS discussions about tracking and monitoring genetic resources, scientists routinely acknowledge genes as information:

> Genetic resources are essentially "packets of informational goods" that are presented as biological material (e.g. an entire specimen, a leaf, a skin, etc.) and include DNA and RNA molecules as well as genes or protein sequences.
>
> (Garrity *et al.* 2009: p. 14)

From manipulating natural information emerges extractable value in, say, new active pharmaceutical ingredients, an industrial enzyme or any of a wide range of biotechnological products. Research can develop and improve on naturally occurring information to generate biotechnologies. The counterfactual Case Studies in this book address access from the starting point that the object is natural information and explore how the particular histories would have unfolded under "bounded openness". They address specific cases of biotechnology R&D over the alkaloid epibatidine (from the poison dart frog – originally found in the tropical Amazon) and the potential implications of IP on diverse compounds from Andean maca.

Many examples of natural information derived products are iconic: one immediately thinks of quinine,[8] aspirin and streptomycin, and more recently Velcro, synthetic vanilla and PCR from *Thermus aquaticus* also come to mind as successful uses of natural information in product development.[9]

Box 1.1 Illustrious antecedents: genetic resources as natural information

The identification of genes as information was intuited even before the discovery of genes. Whether as a metaphor, analogy or homology, scientists have long referred to an informational perspective of life.

In the *Origin of Species* (1859), Charles Darwin understood that there was descent with modification even though he was unaware of the mechanism of transmission. Unbeknownst to Darwin, his contemporary Gregor Mendel, publishes "Experiments on Plant Hybridization" (1865), and gets closer to genes as information when he typifies the "characters" of peas. In "What is Life? (1944)", physicist Edwin Schrodinger is the first to be explicit that "every individual cell, even the most insignificant, must possess a double copy of the code or script." The discovery of deoxyribonucleic acid (DNA) by Watson and Crick (1953) was expressed in the metaphors of information (e.g. a "possible copying mechanism for the genetic material") and later explicitly by Crick (in 1958 in his central dogma of molecular biology, which specifically refers to "genetic information". In the management of genetic resources as natural information, Vogel (1992) fleshes out the homology with artificial information, through information theory. Scientific popularizations have also stressed genes as information. In *River Out of Eden* (1996: p. 5), noted evolutionist Richard Dawkins goes so far as to say that genes are pure information and rivers of information. Unabashedly reductionist, genes as information serves well to develop conceptual and policy approaches.

During the first decade of the CBD, Stone, Swanson and Vogel highlighted the informational dimension of genetic resources, discussing the economic implications in the context of dispersed information and habitat conversion through land use change (Vogel 1992; Stone 1995; Swanson 1997). Vogel took the high road in reduction:

> From biochemistry, we know that genes themselves are nothing more than a sequence of purine and pyrimidine bases bonded on a backbone of phosphate sugar molecules. As a sequence, genes can theoretically be assigned probabilities of occurrence. These probabilities can in turn be quantified as bits of information by the Boltzmann equation of thermodynamics or the equivalent Shannon-Weaver equation of information theory.
>
> (Vogel 1992: p. 14)[10]

Unlike Stone or Swanson who reasoned analogously, Vogel perceived the homology and argued that the economics of information could apply to either artificial or natural information, thereby allowing one to effectively and efficiently capture the value of genetic resources through steps that would align incentives both within and among countries.

Although the homology is the more powerful argument, analogies may be more instructive of the economics. A favorite that appears in various forms in the literature may be helpful: a book (the package) contains a text (the information), which instructs a person upon its reading (the technique).

Barring collectible manuscripts, it is not the physical book where the value lies but the comprehension of the information and its application (Estrella *et al.* 2005: p. 53). For canonical texts, there may be literally millions of sources, from libraries to websites, which have uploaded genomes of model species.[11] The information from a "genetic resource" may be analogously accessed from millions of sources, also including websites given the uploading of sequences. Or at the other extreme, a specimen in a botanical or zoological garden may be the last known copy of what was surely a masterpiece. The sacking of the library at Alexandria is another favorite analogy for extinction (see, for example, "Who Speaks for Earth?" the last episode of Carl Sagan's miniseries *Cosmos*). The Lyceum would have surely housed all the works of Aristotle of which just a quarter survive antiquity.[12] Again, the seeming analogy may be a homology as often the tropical rainforest, home to most of the terrestrial biodiversity, is also literally burned.

Although ABS experts seem to acknowledge the informational dimension of genetic resources, they simultaneously refuse to contemplate the rather obvious implications for policy and legal frameworks. The ability to hold contradictory beliefs and nevertheless be untroubled is what George Orwell famously called "doublethink". Its pervasiveness is stunning and the mere reference to genomics and bioinformatics as if they were based on genetic resources as tangibles for the purposes of policy or regulatory exercises, constitutes a veritable doublespeak.

Doublethink and doublespeak extended to policy and legal frameworks. To date, only one national legal framework makes reference to the informational element in genetic resources. Article 7(i) of the Provisional Measure 2.186-16 of Brazil[13] defines "genetic patrimony" as:

> Information of genetic origin [*natural information?*], contained in samples of specimens of vegetables, fungus, microbes or animals, in whole or part, in the form of molecules and substances derived the metabolism of living or dead organisms found in *in situ* conditions, including domesticated or in *ex situ* conditions, collected from national territory *in situ* sources, in the continental shelf or the exclusive economic zone.[14]
>
> (Italics added)

Disappointingly, the law does not build upon its own definition. The Provisional Measure regulates ABS as if the object of access were still a physical object. It regulates access to the vehicle of natural information or its material support, which invites the accusation of doublethink. Nothing in this law addresses or regulates information through the lens of the economics of information, which justifies time-limited intellectual property rights as a way to recoup the fixed costs of R&D. Seemingly by analogy, but really by homology, patent-like protection over natural information would offset the opportunity costs of habitat conservation.

Costa Rica, Law 7788, on the general conditions for ABS, also includes an interesting reference to information but fails to develop a regulatory framework based on the reference. Orwell is revisited. The law applies to genetic resources and biochemicals, the latter of which are defined as:

> Any material derived from plants, animals, microorganisms or fungus, which contain specific characteristics, special molecules or leads [natural information] to design them.[15]

Though there is an indirect reference to natural information when referring to "leads to design them", emphasis remains on the material component which may be derived therein. The contradiction remains unchallenged.

The sociology of the CBD and national ABS frameworks must be examined to understand how a scientific commonplace – genes, proteins, enzymes and so on, as information – could be so successfully sidelined for such a prolonged period of time, sometimes at the conceptual and academic levels and always in policy and regulatory developments. Chapter 4 offers some tentative explanations and speculations.

Despite the scientific orientation in conservation of biodiversity endeavors, lawyers in particular have been unfazed by the category mistake and actively downplayed the implications of genetic resources as natural information.[16] Narrow interpretations of CBD definitions and, more importantly, reluctance to challenge them have resulted in fairly uniform thinking in ABS, reflected also in existing ABS frameworks. Many who actively participate in the COPs and other ABS forums have been very vocal about defending bilateral contractual approaches to ABS – almost reflexively as if the CBD were not a framework convention which allows for interpretation or amendments in the light of existing evidence or exposure of mistakes.

The summary of the Online Discussion on Article 10 of the Nagoya Protocol, held from March to May 2013, eloquently reflects the tendency to prefer a narrow focus on bilateral ABS contracts as the prevailing instruments to realize the sharing of benefits.[17] Approximately 25 percent of participants in this Online Discussion were lawyers and almost all of them defended a bilateral and contractual approach to ABS – and advocated for PIC and MAT.

Two widely cited publications – *A Guide to the Convention on Biological Diversity* from 1994 and the more recent *An Explanatory Guide to the Nagoya Protocol on Access to Genetic Resources and the Fair and Equitable Sharing of Benefits* from 2013 – go to great lengths in explaining and justifying ABS contracts as the means to ensure PIC, MAT and benefit-sharing over genetic material.[18] Most participants in the ABS debate accept rather uncritically that PIC and MAT will be the means to secure and guarantee the fair and equitable sharing of benefits – through a contract. Advocates for bilateralism downplay the serious limitations of contracts and minimize the fundamental importance of the subject matter which is critical to biotechnologies: natural information.[19]

Modern sequencing technology allows transmission of species in terms of letters of the purine and pyrimidine bases. Access is through a few clicks of the touch pad. The "-omics" revolution (i.e. genomics, proteomics, metabolomics, and so on) and the power of systems biology or "systeomics" (the use of algorithms, equations and modeling of biological processes),[20] are all premised on the reality that one can realize values in genetic resources detached from the vehicle or material support in where they were originally found.[21]

Paul Oldham perceives the pending independence:

[t]rends in the genomic sector suggest a decreasing dependence on physical transfers of biological material and increasing trends towards electronic

transfers because genetic material can be readily expressed as information in the form of A (adenine), G (guanine), C (cytosine) and T (thymine) bases in the case of DNA (deoxyribonucleic acid) and ACG and U (uracil) for RNA (ribonucleic acid). This also extends to amino acids which form the basis of proteins.

(Oldham 2004: p. 13)

In another publication, Oldham also points out that:

The foundations of 'life' are thus coming increasingly into view in a way that may not in fact require acts of collection in countries of origin. Furthermore, the increasing ability to transform biology into an informational good (i.e. DNA and amino acid sequence data) is likely to render physical checkpoints of limited utility in efforts to control such possessions.

(Oldham 2009: p. 17)

The International Civil Society Working Group on Synthetic Biology makes a similar point:

Rather than sourcing genes from nature or gene bank samples, scientists are able to download digital DNA sequences that can rapidly be constructed by commercial foundries. Mail orders of genes and gene samples are now common … As gene synthesis becomes cheaper and faster, it may become easier to synthesize a microbe than to find it in nature or retrieve it from a gene bank.

(International Civil Society Working Group
on Synthetic Biology 2011: p. 32)

In the near future, technologies will be able to mainstream the regular extraction of genetic data from samples without the need for collection, storage and classic taxonomic classification. Advances in DNA barcoding in certain plant species (e.g. through projects such as the International Barcode of Life Initiative – iBOL) and DNA testing kits (e.g. the DNA Explorer Kit or the Biotechnology Forensic DNA Fingerprinting Kit) are already a reality. Identification and certain tests can be undertaken electronically, simply and cheaply through access to the internet and the respective computer program.

Inasmuch as informational constituents can be stripped from their physical medium in biological samples, attempting to institutionalize controls over the flow of information, disembodied at different moments, by different actors, and in different places, is not only impossible but absurd. One can forecast failure with a high degree of certainty. Applying the prevailing logic of ABS focused on the initial physical access to a material entity to biotechnology processes is bound to maintain the current and persistent inequity and unfairness in benefit-sharing.

Ironically, recent papers and documents on the policy and legal aspects of ABS acknowledge the critical relevance and importance of bioinformatics, genomics, etc. and information technologies in R&D in general. More ironically, they do so without any similar acknowledgment of the economic implications for the regulatory frameworks and bilateral approach to ABS.[22] One need only to examine the set of publications by the CBD Secretariat entitled "Bioscience at a Crossroads: Access and Benefit Sharing in a Time of Scientific, Technological and Industry Change." While offering an excellent overview of the state of the art and future direction of science and technology in genetic resources, none of the series of documents – most strikingly the booklet on Industrial Biotechnology – refers to genetic resources as natural information.[23]

The "studied ignorance" was first cited by Oldham *en passant*:

> To date, the implications of these trends ["omics" research] have not been considered in debates surrounding access to genetic resources and benefit sharing in the Convention.
>
> (Oldham 2004: p. 13)

The observation of studied ignorance has entered its second decade and is ever or more relevant. Although "natural information" has been part of ABS literature even before the CBD entered into force in 1993, it has not been part of any of the ABS discussion at the COPs, SBSTTA or related meetings.[24] The point has not been assimilated into the ABS discourse, nor has it been streamlined into policy and legal constructions, for reasons we will examine in Chapter 4.

In the early 1990s, three academics, Christopher Stone, Tim Swanson and Joseph Henry Vogel, independent from one another, realized the power of recognizing genetic resources as information. They reflected on the role of patent-like or *sui generis* systems to capture the value of genetic resources through time-limited monopolies.

Vogel and Swanson were young economists at the beginning of their careers while Stone, in contrast, was already an academic star for his watershed article "Should Trees Have Standing?" (1972). In 1995, he published "What to Do about Biodiversity: Property Rights, Public Goods and the Earth's Biological Riches," which delineates the rationale for conserving biodiversity and genetic resources. Stone separated the intangible content from the genetic material and distinguished between the potential value of a material substance (e.g. a quinine extract from a cinchona tree for further processing) and the potential value of genetic resources as information which can lead to synthesizing or semi-synthesizing useful products (e.g. through use of modern biotechnology and related "omic" disciplines).

Whereas studied ignorance of the works of Vogel and Swanson could be attributed to their young age and lack of stature in the discipline, one cannot so explain the studied ignorance of the work of Stone. Two decades have now passed and time has shown how prescient was the economic approach.

Synthetic biology has proven itself to be a viable industrial and commercial tool, including its different variants in RNA synthetic biology, cyanobacterial synthetic biology and plant synthetic biology.[25] Its advent complements organic chemistry which is focused on analysis and synthesis of chemical compounds, including the subdiscipline medicinal chemistry.[26] Classic examples of medicinal chemistry products include cholesterol-lowering statins (atorvastatin), the chemotherapeutics vinblastine/vincristine and the analgesic ziconotide (Prialt). All were derived or inspired by naturally occurring metabolites found in plants, animals and microorganisms, illustrating the link between organic and natural product chemistry.

As Stone, Swanson and Vogel stressed, encoded natural information presents certain features of a public good. It is non-rivalrous and can be non-exclusionary, depending on the policy option and legal architecture designed to protect it. The analogy of books can be extended to CDs. Listening to the music on a CD does not affect or limit the possibilities of multiple listeners and readers enjoying exactly the same content. The justification is to allow financial viability for innovation and creativity. The IP system plays a critical role in providing such protection to innovators and creators.[27]

Stone's sweeping analysis complements a more detailed proposal by Vogel regarding how to specifically regulate genetic resources – as informational assets[28] – and prevent habitat loss. Three years before Stone published his article "What to Do about Biodiversity," Vogel published *Privatisation as a Conservation Policy* (Vogel 1992) as a monograph from the think tank CIRCIT in Australia, which was subsequently released by Oxford University Press as *Genes for Sale* (1994). The book explicitly advocates a form of "equal protection of artificial and natural information" and stressed that value lies in natural information and not the physical, tangible vehicle of that information (the specimen, sample, leaf, stem, etc.).

Key to the argument was that genetically coded functions lie at all levels from those restricted to individuals to perhaps all individuals of a species from different families. Hence, many countries are sources of *the same* natural information through its dissemination across taxa and jurisdictions. In this context the notion of "country of origin" as the sole beneficiary through bilateral ABS contracts, proposed by the CBD, is intrinsically unfair and inequitable.[29] Why should one country take advantage of or benefit exclusively from natural information which neighboring countries also possess and conserve?

The notion of diffusion of natural information across jurisdictions is an empirical issue. In his work on biodiversity in the patents system, Oldham and his colleagues make a very strong case regarding shared and common species used in patented products:

> In some cases species may be unique to a particular country. However, species do not respect political boundaries and are frequently distributed in more than one country … the bulk of patent activity is concentrated around a small number of well-known and cosmopolitan species … *Species that are limited to one or a very small number of countries are likely, on the basis of available distribution data, to be exceptions rather than the rule.* [30]

(Italics added)

What this means in simple terms is that if natural information from species A is used in R&D, it is highly likely that species A is also present in other countries. Synthesizing Oldham's empiricism with Vogel's point about diffusion of natural

Table 1.2 An initial look at bounded openness

Natural information	Object of access (without bounds)
⇩	Bounds imposed upon utilization
Disclosure without specification in procurement of intellectual property	
⇩	Benefit-sharing among countries of origin
Monopoly-limited in time: fixed royalty rates dependent on features of good or service (generally a biotechnology product)	

information across taxa, the number of source species can even go to ubiquity, as is the case with ATP synthase.

The country of origin of the physical sample or biological resource is largely irrelevant for the sharing of benefits. Its only real relevance is for facilitating the determination of the geographic diffusion of the natural information. As the years and decades pass, more and more observers and experts in ABS discussions have come to recognize that shared genetic resources "pose a problem" in terms of fairness and equity yet few seem willing to even entertain the known explanations from the economics of information. Transboundary natural information is the rule and not the exception in terms of R&D. However, even the Nagoya Protocol, which entered into force in October 2014, gets this the other way round. For the Protocol, shared genetic resources are assumed to be the exception in the context of ABS!

In this context and citing Theodosius Dobzhansky's famous "Nothing in Biology Makes Sense Except in the Light of Evolution" (1973), Vogel quips that "Nothing in ABS Makes Sense Except in the Light of the Economics of Information."[31]

The alternative conceptual framework for an ABS approach based on genetic resources as natural information means a multilateral non-contractual regime focused on fairness and equity in the distribution of monetary benefits. Vogel calls it "bounded openness", which is the generic term coined by the political scientist Chris May (2010). Existing permits, authorizations and collaborative arrangements may continue to exist, but for the future monetary benefits, the "bounded openness" based regime would enter into play (Table 1.2). Chapter 5 describes in detail the nature of the proposed regime.

The most important benefit to be shared among Providers would be money, divisible once a product is developed and successfully commercialized through intellectual property protection. These could then offset opportunity costs of habitat conservation. This is a key difference with the bilateral approach. Analysis over the past two decades has focused substantially on the operational aspects of ABS but hardly any reflection has been given to how ABS can relate directly to and align incentives for the conservation of transboundary genetic resources. The bilateral approach focuses on defining who is the national ABS authority, what are the administrative steps for securing access, what is the substantial content of contracts, and so on. Because the three objectives of the CBD, conservation, sustainable use and benefit-sharing, are inexorably intertwined, the low benefits of the bilateral approach go against the first two objectives.

The approach is pragmatic and helps a market emerge for conservation. Vogel stresses that estimates of future royalties not be folded into any calculation of total economic value (TEV) of biodiversity for cost-benefit analysis and argues that "any such calculation is a meaningless number."[32] He further concludes that:

> People should pay, not because habitats must compete with timber, cattle, and dams, but because there is tremendous political pressure by the vested interests behind timber, cattle, and dams to encroach on protected habitats. The generation of revenues from the sustainable use of biological diversity can create countervailing pressures against the exterminators. This has been the humble alternative to [what E.O. Wilson has called] "bankrupt economics".
>
> (Vogel 1997: p. 8)

Other scholars, including Peter Drahos and Gavin Stenton, have also noted the meager monetary benefits negotiated under the existing ABS models, even in industrialized countries such as Australia – despite the near trillion dollar per year world markets that derive from the genetic resources industry.[33]

And the non-monetary benefits? Economist are biased toward money because non-monetary benefits are not divisible, their value is hard to estimate and may also not reflect the highest social return on investment. Some of these benefits may already be part of regular practices within research communities and include transfer of know-how, co-publications, development of training programs, sponsoring internships and thesis work.[34] Although valuing non-monetary benefits can remain part of the new ABS architecture, the attention heaped on non-monetary benefits distracts from what will always have to be the lion's share of benefits: money.

Money to be paid for some future event is hard to comprehend. As noted from behavioral economics and "prospect theory", we should not confuse the mathematical expectations of a low probability and high payoff event with the actual high payoff of the event. In other words, there may be very few blockbusters like PCR, but when they happen (and they will), the expectation is most significant. Hence, bounded openness allows us to cast a large net.

An alternative ABS framework based on natural information distances itself from the narrow interpretation of sovereignty by the CBD and its corollaries in PIC and MAT. Throughout the 1990s and new millennium, Vogel strongly advocated reform within the CBD process, reminding audiences of its status as a framework treaty. Lack of success in persuading the COPs has led him to theorize on the role of scholarship in policymaking, where he perceives a "Tragedy of Unpersuasive Power" for which the CBD becomes exemplary (Vogel 2013). Other explanations are also suggested in Chapter 4.

Almost like a controlled experiment in studied ignorance, another economist, Timothy Swanson, converged on the same conclusions as Vogel and Stone. He asserted in *Global Action for Biodiversity* that "the value of biodiversity lies in its informational content," adding to the few voices which departed from the enthusiasts of the bilateral ABS approach. Swanson argued that nature is an important source of information that contributes to industries and daily lives in myriad forms. He conceptualized the need for an "informational resources right system" which he called a Protocol for Registration of Sui Generis Rights in Natural Resource Information (Swanson 1997). His proposal has also not found resonance in the COPs and the CBD process.

The recognition of the utilization of genetic resources as the utilization of natural information will require a dramatic "paradigm shift" for CBD policymaking. What is both bizarre and embarrassing for the ABS community is that the paradigm shift takes us to what is "normal science" in biology. Persistence in treating the object of access as material is no longer tenable and, at the very least, misplaced.

NOTES

1 Although scholars debate whether the CBD is a "framework" or an "umbrella" convention, this book takes the former position. It establishes certain principles and determines general compromises by countries but does not include *specific* ABS obligations and qualifies substantially almost every provision, thereby allowing interpretation for countries to specify and develop national laws and regulations.
2 Graham Dutfield, Professor of International Governance at the University of Leeds, has specialized in ABS and IP and considers genetic resources to be "things". They are in his view a tangible package with an informational content (email communication, August 7, 2013).
3 Almost all laws and regulations have referred to access to genetic resources and derivatives or derived products as material or tangibles. The Nagoya Protocol is the exception as it has developed provisions which place the *utilization* of genetic resources under its scope. Nevertheless, laws and regulations worldwide do not adopt the "utilization" dimension that appears in the objective of the CBD (Article 3).
4 The first legal ABS framework to include "derivatives" or "derived products" within its scope was Decision 391 of the Andean Community on a Common Regime on Access to Genetic Resources (1996). The Andean Community is an economic integration bloc created in 1969 and formed by Bolivia, Colombia, Ecuador and Peru. Article 1 of Decision 391 defines "derived product" as a "Molecule, combination or mixture of natural molecules, including crude extracts of living or dead organisms or biological origin, derived from the metabolism of living things." Other national ABS frameworks have similar definitions to broaden their scope and coverage. The emphasis, however, is always on the material dimension of genetic resources. For a detailed analysis of Decision 391, see Ruiz, M. (2008) *Guía Explicativa de la Decisión 391 y una Propuesta Alternativa para Regular el Acceso a los Recursos Genéticos en la Región Andina.* GTZ, SPDA, The MacArthur Foundation, Lima, Peru.
5 Vehicles are also valuable when used as commodities or directly in bulk quantities, directly processed as natural products. This is often the case in biotrade activities. As long as no IP is attached to the value added, the vehicle should not be an object of access.
6 Developing countries realized early in the ABS discussions that focusing regulations on specific genes limited their capacity to control access to and use of other valuable components of biodiversity thereby forgoing participation in benefit-sharing schemes and limiting their sovereignty. Hence "derivatives" or "derived products" expands coverage of ABS frameworks and reaffirms sovereignty over a broader set of biodiversity components.

Under the expansion, biotrade related activities and enterprises, or distinct phases therein, may fall within the scope of ABS. The Biotrade Principles and Criteria (adopted in 2007) include a specific reference to [access and] benefit-sharing under Principle 3, "Fair and equitable sharing of benefits derived from the use of biodiversity." These principles broaden the scope of [access and] benefit-sharing even more by applying to "the use of biodiversity." The GIZ sponsored ABS Initiative, the Union for Ethical Biotrade and the United Nations Conference on Trade and Development (UNCTAD) have tried to define the implications of ABS in the context of biotrade related activities. See, for example, Biotrade Initiative (2000) *UNCTAD Biotrade: Some Considerations on Access, Benefit Sharing and Traditional Knowledge.* Working Paper. Prepared for the UNCTAD Expert Meeting on Systems and National Experiences for Protecting Traditional Knowledge, Innovations and Practices. Geneva, October 30, 2000. Available at www.biotrade.org/ResourcesPublications/Some%20considerations%20on%20 ABS%20and%20TK.pdf

7   EPO Board of Appeals Decisions, case V0008/94 – Opposition Division, December 8, 1994.

8   Though once dismissed as a "long shot" in the mid-1990s, synthetic products have resurfaced as the focus of attention, given their potential significance in biosafety, health, agriculture and a wide range of industries. During COP 12 in Korea (October 2014), some organizations called for the adoption of international rules to regulate synthetic biology. The ETC Group released: *Regulate Synthetic Biology Now: 194 Countries* (October 19, 2014), which presents the environmental, ethical and technical justifications for regulation. See The International Civil Society Working Group on Synthetic Biology. *A Submission to the Convention on Biological Diversity's Subsidiary Body on Scientific, Technical and Technological Advice (SBSTTA) on the Potential Impacts of Synthetic Biology on the Conservation of Biodiversity.* October 17, 2011. Available at www.cbd.int/doc/emerging-issues/Int-Civil-Soc-WG-Synthetic-Biology-2011-013-en.pdf

9   Iconic examples abound: streptomycin, indicated for tuberculosis, was developed from the microorganism actinobacterium (*Streptomyces griseus*) in the 1940s; quinine from the bark of the cinchona tree, used to treat malaria and discovered in the fifteenth century, was chemically synthesized into more potent drugs such as primaquine and chloroquine; vinblastine and vincristine were developed from the rosy periwinkle (*Catharanthus roseus*); Prialt –a potent analgesic – was derived from *Conus magus*, a sea cone. Taxol (a drug derived from *Taxus brevifolia*) is especially interesting because it offers an example of the utilization of natural information where the biological material is still important in the development process. See Walsh, V. and Goodman, J. (1999) "Cancer Chemotherapy, Biodiversity, Public and Private Property: The Case of the Anti-Cancer Drug Taxol." *Social Science & Medicine* 49: 1251–1255.

10   What the Boltzmann equation of statistical entropy or the isomorphic Shannon equation for entropy mean is that the sequence of bases in DNA is literally *bits* of information. Hence, DNA and a book are homologous. Seen thus, one can argue for an *equal* legal protection of natural and artificial information for the DNA and the book, respectively.

11   See example of the naked mole rat (*Heterocephalus glaber*) at www.naked-mole-rat.org/

12   See Lukács, B. *A Note to the Lost Books of Aristotle*, www.rmki.kfki.hu/~lukacs/ARISTO3.htm

13   Provisional Measure 2.186-16 of Brazil, which regulates paragraph II, numerals 1 and 4 of Article 225 of the Constitution, Article 8(j), Article 10(c) and Articles 15 and 16 of the Convention on Biological Diversity, provisions on the genetic patrimony, protection and access to associated traditional knowledge, benefit-sharing and access to technology for conservation, use and its transfer, among others (August 23, 2001). Available at www.farmersrights.org/pdf/americas/Brazil/Brazil-access01.pdf

14 Author's translation of Article 7.I.

15 Author's translation of Article 7 (Definitions) in Law 7788 on Biodiversity (April 23, 1998).

16 As a lawyer, I have nothing against my own profession. However, I do feel that the CBD process and ABS in particular have been excessively legalistic and complex, in contrast to the simplicity of the synthesis of biology and economics. At least comparatively. Closer attention by lawyers to that synthesis could have resulted in more operational ABS regimes, both internationally and nationally, in the early years of the CBD.

17 The Online Discussion was convened by the CBD Secretariat from April to May 2013 as a follow-up to the Reflection Meeting on the Global Multilateral Benefit Sharing Mechanism, convened by the Fridtjof Nansen Institute and the ABS Capacity Development Initiative sponsored by GIZ (see FNI Report 10/2011, available at www.fni.no/doc&pdf/FNI-R1011.pdf). The official synthesis of this Online Discussion is available at www.cbd.int/doc/?meeting=ABSEM-A10-01

18 Glowka, L., Burhenne-Guilmin, F. and Synge, H. (1994) *A Guide to the Convention on Biological Diversity*. Gland and Cambridge: IUCN; and Greiber, T., Peña Moreno, S., Ahren, M., Nieto Carrasco, J., Chege Kamau, E., Cabrera, J., Olivia, M.J. and Perron-Welch, F. (2013) *An Explanatory Guide to the Nagoya Protocol on Access to Genetic Resources and the Fair and Equitable Sharing of Benefits*. Gland, Switzerland: IUCN.

19 Many stakeholders only ask how to develop PIC and MAT contents without questioning whether PIC and MAT would achieve fairness and equity, in both their horizontal and vertical dimensions of benefit sharing. Horizontal refers to benefit-sharing between countries that may share similar biodiversity components; vertical refers to participating in the R&D process as value is added to biodiversity components. See Winter, G. (2009) "Towards Regional Common Pools of GRs – Improving Equity and Fairness in ABS," in Kamau E.C., Winter, G. (2009) (eds.) *Genetic Resources, Traditional Knowledge and the Law. Solutions for Access and Benefit Sharing*. London: Earthscan.

20 For a simple and rapid introduction to systems biology, see www.sysbio.de/info/background/WhatIs.shtml

21 A clarification is warranted. I am *not* suggesting that the material, physical support, is unimportant but only pointing out that *for the specific purpose of sharing monetary benefits from the value added through intellectual property to provide incentives for conservation*, one must take a reductionist view that enables the development of an effective benefit-sharing mechanism, namely, bounded openness, as will be explained in Chapter 5.

22 For a description of the complexities in R&D that utilizes genetic resources, i.e. natural information, and the policy and legal challenges in the context of the Nagoya Protocol, see Pastor, S. and Ruiz, M. (2008) *The Development of an International Regime on Access to Genetic Resources and Fair and Equitable Benefit Sharing in a Context of New Technological Developments*. Initiative for the Prevention of Biopiracy. SPDA. Lima, Peru. Year 4, No. 10, April 2008. Available at www.biopirateria.org/documentos/Serie%20Iniciativa%2010.pdf

23 Available at www.cbd.int/abs

24 During COP 12 in Pyeongchang, Korea, a side event organized by the Woodrow Wilson Center presented the challenges posed by synthetic biology in the context of ABS.

25 The pharmaceutical company Sanofi announced in 2013 the launch of a semi-synthetic antimalarial drug derived from artemisinin, based on R&D conducted at UC-Berkeley. See http://newscenter.berkeley.edu/2013/04/11/launch-of-antimalarial-drug-a-triumph-for-uc-berkeley-synthetic-biology/

26 For a review of the challenges and potential of synthetic biology, see Luo, Y. *et al.* (2012) "Challenges and Opportunities in Synthetic Biology for Chemical Engineers." *Chemical Engineering Science* 103: 115–119. Available at http://dx.doi.org/10.1016/j.ces.2012.06.013

27   Many academics perceive that patents encumber R&D. Despite the moral or ethical positions in regard to patents and plant breeders' rights, they continue to be a prevailing tool to protect innovations including biotechnology. Activists and influential organizations such as GRAIN, ETC Group and TWN question the moral and ethical grounds for patenting life forms. Noted scholars such as James Boyle, Paul Ulhir and Jerome Reichmann have warned against "enclosure" and the "highly restrictive environments" fostered by IP since the early 1990s. See, for example, Boyle, J. (2003). "The Second Enclosure Movement and the Construction of the Public Domain." *Law and Contemporary Problems* 66: 33–74.

28   Vogel and Stone arrived at essentially the same conclusion because it was low-hanging fruit. Stone's article appears in an issue dated 1994–1995, which means a release date of 1995. Vogel's book from CIRCIT has an ISBN number and would establish the chronology as 1992, specifically as a special limited edition released on November 16, 1992 for the AIC Conference on Biodiversity held in Sydney, Australia. Oxford University Press republished the book, whose contents were only slightly modified, under the title *Genes for Sale* in 1994, *before* Stone's publication. Personal communication, April 9, 2014.

29   A country of origin of genetic resources is defined in the CBD as the "country which possesses those genetic resources in *insitu* conditions." Although it is physically impossible that two or more countries simultaneously possess the same matter, it is not only possible but the norm that two or more countries possess the species that hold the same natural information.

30   Oldham, P., Hall, S. and Forero, O. (2013) "Biological Diversity in the Patent System." *PLoS ONE* 8(11): 6.

31   See Vogel, J.H. (2008) "Nothing in Bioprospecting Makes Sense Except in the Light of Economics," in Sunderland, N., Graham, P., Isaacs, P., McKenna, B (eds.) *Toward Humane Technologies: Biotechnology, New Media and Ethics*, 65–74. Rotterdam: Sense Publishers Series.

32   For an analysis of this argument, see Vogel, J.H. (1997) "White Paper: The Successful Use of Economic Instruments to Foster the Sustainable Use of Biodiversity: Six Cases from Latin America and the Caribbean," *Biopolicy Journal* 2, (Paper 5) (PY97005). Archived with the British Library in hard copy, ISSN 1363-2450. Available in English, Spanish and Portuguese. www.bioline.org.br/request?py97005

33   The most publicized example of a bioprospecting model (the National Biodiversity Institute – INBIO – in Costa Rica) has also drawn criticism:

> The model "example" of such an initiative [bioprospecting] is that of the major pharmaceutical company Merck. In 1991, a year in which the firm made profits of US$8.6 billion, a contract was signed with Costa Rica, home to between 5% and 7% of the world's species. In exchange for exclusive rights to screen, develop and patent new products from plants, microorganisms and animals, a total of US$1.1 million was paid to a local biodiversity program [INBIO] and the National Environment Ministry ... this represents a fee of US$2 per specie and on a global scale means that such a rate of exchange of the world's genetic resources could be purchased for US$20 million!

Quote from: Stenton, G. (2003) "Biopiracy within the Pharmaceutical Industry: A Stark Illustration of Just How Abusive, Manipulative and Perverse the Patenting Process can be Towards Countries of the South." *Hertfordshire Law Journal* 1(2): 30–47, at p. 42.

The influence of INBIO resonates strongly to date. For years, countries have sought to set up similar models, under bilateral ABS frameworks. The collapse of INBIO is a cautionary tale.

34   The Bonn Guidelines on ABS (Decision VI/24, COP 6, 2004) include an Appendix II with a list of the different forms of monetary and non-monetary benefits. See www.cbd.int/decision/cop/default.shtml?id=7198

Chapter 2

Regulatory trends in ABS

Secrecy as the enabler of the bilateral model

Articles 1, 15, 16 and 19 of the CBD establish minimum ABS and related principles and standards for Contracting Parties, which become operational upon incorporation into national regulatory frameworks. A large and widely cited literature exists which analyzes these articles and indicates that Contracting Parties have flexibility in developing ABS administrative procedures within basic legal contours (Glowka *et al.* 1994; Glowka 1998; Nnadozie *et al.* 2003; McManis 2007; Kamau and Winter 2013a; Suneetha and Pisupati 2009). Over time, the contours themselves have become the issue.

All existing national ABS frameworks recognize the right of the State, through a national competent authority, to authorize and grant access to genetic resources. The right is viewed as almost synonymous with sovereignty, as commonly understood. Most procedures feature the following sequence: presentation of an access application; review by a competent authority and, if successful, the authorization and granting of PIC; negotiation between the national authority and the user of the genetic resource(s); the signing of an access contract, according to and reflecting MAT; and transmittal of official approval.

In some cases, intermediate or parallel procedures involve permits or authorizations for access to *biological* samples, and access to the TK of local and indigenous communities which in turn also require PIC and MAT negotiations. The PIC from local and indigenous communities is a complex issue leading invariably to issues of self-determination and rights regarding territoriality, consultation and participation.[1]

Prior informed consent and MAT are at the heart of all regulatory frameworks. No better example can be found than the Andean Community Decision 391, a pioneering and still influential legal framework. At least a half dozen different contractual instruments form part of the Andean ABS regulation, making the transaction costs *à la carte*.[2] Concerns since enactment of Decision 391 and the Philippines Executive Order 247 have focused on the stifling effects of ABS frameworks on research in a broad range of fields, from the basic science of taxonomy, to highly specific pharmaceutical research and the development of biological control agents.[3]

Costa Rica is often presented as evidence that even complex ABS frameworks – where agreements, bioprospecting contracts, permits and concessions coexist – need not stifle research endeavors.[4] The country is a good example of a commendable and positive scientific program through the National Biodiversity Institute (INBIO) albeit operating under a questionable policy and regulatory approach in terms of equity and fairness in monetary benefit distribution. This will be further elaborated in Chapter 3.

The law and regulation in Costa Rica were specifically designed for INBIO, making it virtually the only privileged provider of genetic resources![5] Brazil and its Provisional Measure 2.186-16/2001 regarding access to the genetic patrimony of Brazil could also be perceived as an up and coming success in implementation of ABS principles as judged by the number of authorizations issued and agreements concluded.[6] Likewise, Australia has also been heralded as an example of ABS requirements resulting in numerous ABS agreements concluded.

If the quantitative dimension is a measure of success, some countries have been more successful than others in concluding ABS agreements. The mere number of ABS agreements is not in and of itself a measurement of success, as the qualifiers "fair and equitable" cannot be assessed in any of the concluded agreements. Evidence that the hurdles are significant is the international pressure and national regulatory efforts to exempt basic science, taxonomic and human pathogen research and barring that, allow "non-commercial" research agreements.

In most circumstances, however, due to the national bureaucratic hurdles that ABS frameworks have imposed, international concerns arise over the effects of ABS on, particularly, basic science and taxonomic research and hence the advocacy of "non-commercial" ABS research agreements.[7] The Nagoya Protocol does include a provision to further facilitate access for non-commercial research which would cover research for species classification, population distribution studies and systematics.[8]

To further streamline the approval process, the "bilateral approach" has repeatedly advocated model ABS contractual clauses assuming that the approach itself is non-negotiable.[9] In other words, negotiations will only be between Provider and a User through an access agreement which incorporates provisions for a fair and equitable sharing of benefits. The Provider is defined as a country of origin, local or indigenous community or institution and the User, a person, research institution, company or university. As Gudrun Henne pointed out early in ABS discussions, PIC and MAT underpin both the administrative procedure and the ABS agreements (Table 2.1).[10]

The approach is vigorously defended across diverse stakeholders, including legal experts, delegates from both Provider and User countries, and representatives of traditional and indigenous communities.[11] Most participants in COP and scholars discussing how to make benefit-sharing effective support bilateralism as an expression of sovereignty and reiterate the correctness of PIC and MAT provisions on ABS in the CBD. They assume that the equity and fairness in the sharing of benefits from the utilization of genetic resources is merely a reflection of the quality

Table 2.1 The contractual approach in the CBD: relevant provisions (italics inserted for the relevant language)

Article 1 (Objectives)	The objectives of this Convention, to be pursued in accordance with its relevant provisions, are the conservation of biological diversity, the sustainable use of its components and the fair and equitable sharing of the benefits arising out of the utilization of genetic resources, including by appropriate access to genetic resources and by appropriate transfer of relevant technologies, taking into account all rights over those resources and to technologies, and by appropriate funding.
Article 15 (Access to Genetic Resources)	1. Recognizing the *sovereign rights* of States over their natural resources, the authority to determine access to genetic resources rests with the national governments and is subject to national legislation.
	2. Each Contracting Party shall endeavor to create conditions to facilitate access to genetic resources for environmentally sound uses by other Contracting Parties and not to impose restrictions that run counter to the objectives of this Convention.
	3. For the purpose of this Convention, the genetic resources being provided by a Contracting Party, as referred to in this Article and Articles 16 and 19, are only those that are provided by Contracting Parties that are *countries of origin* of such resources or by the Parties that have acquired the genetic resources in accordance with this Convention.
	4. Access, where granted, shall be on *mutually agreed terms* and subject to the provisions of this Article.
	5. Access to genetic resources shall be subject to *prior informed consent* of the Contracting Party providing such resources, unless otherwise determined by that Party.
	6. Each Contracting Party shall endeavor to develop and carry out scientific research based on genetic resources provided by other Contracting Parties with the full participation of, and where possible in, such Contracting Parties.
	7. Each Contracting Party shall take legislative, administrative or policy measures, as appropriate, and in accordance with Articles 16 and 19 and, where necessary, through the financial mechanism established by Articles 20 and 21 with the aim of sharing in a fair and equitable way the results of research and development and the benefits arising from the commercial and other utilization of genetic resources with the Contracting Party providing such resources. Such sharing shall be upon *mutually agreed terms*.

(continued)

Table 2.1 (cont.)

Article 16 (Access to and Transfer of Technology)	1. Each Contracting Party, recognizing that technology includes biotechnology, and that both access to and transfer of technology among Contracting Parties are essential elements for the attainment of the objectives of this Convention, undertakes subject to the provisions of this Article to provide and/or facilitate access for and transfer to other Contracting Parties of technologies that are relevant to the conservation and sustainable use of biological diversity or make use of genetic resources and do not cause significant damage to the environment.
	2. Access to and transfer of technology referred to in paragraph 1 above to developing countries shall be provided and/or facilitated under fair and most favorable terms, including on concessional and preferential *terms* where *mutually agreed*, and, where necessary, in accordance with the financial mechanism established by Articles 20 and 21. In the case of technology subject to patents and other intellectual property rights, such access and transfer shall be provided on terms which recognize and are consistent with the adequate and effective protection of intellectual property rights. The application of this paragraph shall be consistent with paragraphs 3, 4 and 5 below.
	3. Each Contracting Party shall take legislative, administrative or policy measures, as appropriate, with the aim that Contracting Parties, in particular those that are developing countries, which provide genetic resources are provided access to and transfer of technology which makes use of those resources, on *mutually agreed terms*, including technology protected by patents and other intellectual property rights, where necessary, through the provisions of Articles 20 and 21 and in accordance with international law and consistent with paragraphs 4 and 5 below.
	4. Each Contracting Party shall take legislative, administrative or policy measures, as appropriate, with the aim that the private sector facilitates access to, joint development and transfer of technology referred to in paragraph 1 above for the benefit of both governmental institutions and the private sector of developing countries and in this regard shall abide by the obligations included in paragraphs 1, 2 and 3 above.
	5. The Contracting Parties, recognizing that patents and other intellectual property rights may have an influence on the implementation of this Convention, shall cooperate in this regard subject to national legislation and international law in order to ensure that such rights are supportive of and do not run counter to its objectives.
Article 19 (Handling of Biotechnology and Sharing of its Benefits)	1. Each Contracting Party shall take legislative, administrative or policy measures, as appropriate, to provide for the effective participation in biotechnological research activities by those Contracting Parties, especially developing countries, which provide the genetic resources for such research, and where feasible in such Contracting Parties.
	2. Each Contracting Party shall take all practicable measures to promote and advance priority access on a fair and equitable basis by Contracting Parties, especially developing countries, to the results and benefits arising from biotechnologies based upon genetic resources provided.

of the contract and its negotiation. A quick overview of international, national and regional ABS instruments confirms that contracts are the preferred tool under which Users and Providers legally relate.[12]

Some literature on ABS contracts suggests criteria to negotiate contracts and includes recommendations on the need to consider the markets for genetic resources, the use of TK, expectations and interests of the parties involved.[13] The questions asked are: are market values known? And is TK being used in any form?

Over a period of more than two decades, relatively few bioprospecting projects exist where PIC and MAT correspond to the contours envisioned by the CBD and national ABS frameworks. Of the few which do exist, none has been a success and all could, I daresay, be classified as failures in terms of fairness and equity. If we use ten Kate and Laird's estimates of per annum sales in the global markets in genetic resources (US$500–800 billion) (ten Kate and Laird 1999: p. 1), a distribution question emerges: what share should go to countries of origin under ABS? More recent and similar studies refer to a multi-billion dollar sector of the global economy.

How much R&D has conformed to ABS principles and frameworks and benefited Providers?[14]

The recent works of Paul Oldham and Edward Hammond have examined the patent system and discovered thousands of patents related to genetic resources.[15] One conclusion from these global assessments is that access to genetic resources has overwhelmingly taken place without PIC or MAT, and thereby in the margins of legality. Furthermore, these sources and the work by David Newman and Gordon Cragg regarding the percentage of new active compounds being admitted by the US Food and Drug Administration (FDA) which are natural products or natural product derivatives, indicate vast amounts of natural information being extracted and derived from a wide range of genetic resources and thereafter used in R&D.[16]

Given that there is almost no access with PIC and MAT but there seems to be abundant *use* of natural information in R&D, only two explanations exist: either misappropriation or biopiracy in the sense of non-authorized access to *in situ* genetic resources is taking place on a massive scale or the use of natural information for R&D can be exercised without accessing *in situ* (or even *ex situ*) collected genetic resources through digital and electronic means.

The data from Oldham, Hammond, Newman and Cragg indeed indicate the existence of a more "sophisticated" type of bioprospecting – *digital* prospecting which returns us to the category mistake of Article 2 of the CBD that defines "genetic resources" as "material".[17]

The extent of the problem becomes clear when bioprospecting is defined in reductionist terms: the systematic search for natural information useful for R&D. Bioprospecting is a tool that facilitates initial access to natural information which may then be disembodied into a "virtual world". Its transformation does not negate the ABS obligation for benefit-sharing. Evaluation of the performance of the small percentage of genetic resources which have been accessed with PIC and MAT cannot meet any reasonable understanding of what is "fair and equitable" as only one Provider receives a benefit for what is almost always a shared or transboundary resource. However, blame should not be assigned to Contracting Parties alone.

Searching for "successful" ABS case studies – involving bioprospecting – is like searching for the perpetual motion machine. But if expectations are purposefully lowered, trumpeting royalties of 1 percent or less for just one Provider, then failure can be defined as success. The accusation of Orwellian doublespeak has traction. Bilateral contracts (PIC and MAT) as envisioned by the CBD and all national frameworks are condemned to fail because only one country can get benefits even if various countries have the same resource (natural information). Microeconomics imply that the country of "extraction" – let's dispense with "country of origin" as it assumes endemism – will only get a royalty rate which reflects the marginal cost of extraction. Even in the unusual case where the natural information is unique to one country, other natural information from more cosmopolitan species is not unique and the competitive process for the latter pulls down the benefits for the former. In other words, incentives would emerge to bioprospect the least threatened species thereby frustrating the first two objectives of the CBD, conservation and sustainable use.

Failure also includes many initiatives and projects which have not been concluded or have had to be suspended, as the cases in Table 2.2 indicate, due to administrative hurdles and often State inaction.

Ultimately, the most egregious failure is the knowledge that bioprospecting and R&D are being undertaken with extensive use of natural information and blatant disregard for the existing ABS frameworks and mechanisms, plus a systematic inability to capture benefits. Non-enforcement appears willful.

Nevertheless, advocates of bilateralism are unfazed and have gone to tremendous lengths to facilitate and streamline contractual approaches. Almost no one questions the meaning and implications of fairness, equity and sovereignty over genetic resources in the mandates of the CBD. Invariably, training courses in ABS involve "capacity building in ABS contracts negotiations" which is basically how to conclude ABS contracts, a Sisyphean task.[18]

Sovereignty is commonly understood as the authorization of the State in determining the conditions which need to be met and expressly agreed upon in order to obtain genetic resources found in its territory or under its jurisdiction. As Table 2.3 shows, these conditions are usually set in existing ABS legislation and guide the drafting of access contracts or MTAs. The text of the CBD leaves little room for alternatives to bilateral negotiation of contracts although national administrative procedures under which they take place may provide some leeway and flexibility. For this reason, the Nagoya Protocol is both significant and hopeful as it suggests timidly, and subject to various qualifiers, a possible non-contractual approach and multilateralism in Articles 10 and 11. Both the capacious language and the contradictions have consequences.

Delegates are often befuddled by the ambiguous and circumlocutory language of the CBD and Nagoya Protocol. To address a very real need, various guidelines have emerged. The IUCN Explanatory Guide to the CBD (Glowka *et al.* 1994) has been particularly influential and required reading for anyone seeking to

Table 2.2 Examples of failed bioprospecting projects

Project	Country	Year	Reasons for failure (interruption of projects)
Bioandes and Andes Pharmaceuticals (US) – bioprospecting project in search of active components in medicinal plants (genetic resources and derivatives) in the national parks system[19]	Colombia	1997–1998	National competent authority (Ministry of the Environment) denied the company access to Colombian genetic resources on the grounds of insufficient benefits being provided
ICBG – Peru – Bioprospecting project in search of active compounds from medicinal plants form indigenous communities[20]	Peru	1993–2000	Final collecting permits were not issued by the national authority at the time (the National Institute for Natural resources – INRENA), 4 years after university responsible for collecting petitioned
ICBG – Maya/Mexico – Bioprospecting project in the Chiapas region in search of medicinal compounds from plants[21]	Mexico	1999–2000	Suspended due to disputes with Mayan indigenous peoples regarding PIC, benefit-sharing, land rights and TK
Biozulua Project – project to document TK related to medicinal plants of indigenous communities[22]	Venezuela	2000–2002	Suspension by the Ministry of the Environment due to lack of PIC from indigenous communities for the use of TK
Pro Benefit Project – project sponsored by the GIZ to support a PIC/MAT process to initiate bioprospecting activities[23]	Ecuador	2003–2007	The Ministry of the Environment was unable to provide and formalize PIC/MAT and conclude an access contract with partners, who in turn decided to terminate the project on amicable terms

understand the CBD and ABS. More recently, the IUCN Explanatory Guide to the Nagoya Protocol (Greiber *et al.* 2013) follows the same approach and style in interpreting the Protocol's provisions on ABS. The influence of the latter remains to be seen.

Given their strict legal perspective, neither guide questions the fairness and equity of bilaterally negotiated contracts, much less their efficiency. They focus instead on how to best achieve PIC from a national authority or a community or a Provider in general and negotiate MAT with a Provider.

Access agreements or contracts have become the primary means for Parties not only to authorize access to genetic resources, but also to agree on the return of benefits from subsequent use. Indeed, it may prove difficult to negotiate benefit-sharing independently of or after negotiating an access agreement.[24]

Table 2.3 ABS laws and regulations worldwide

Country	Procedure	Enabling bilateral instrument	Enabling confidentiality provisions
Bolivia	Access application – ABS contract negotiation – contract approval	Access contract Accessory contract	Yes
Brazil	Access application – ABS contract negotiation – contract approval	Access contract	No
Costa Rica	Access application – ABS contract negotiation – contract approval	Bioprospecting and basic research agreement	No: royalties are fixed in the ABS regulation; but there is the possibility to request confidentiality in regard to IP conditions
Colombia	Access application – ABS contract negotiation – contract approval	Access agreement	Yes
Panama	Access application – ABS contract negotiation – contract approval	Access agreement	Yes
Peru	Access application – ABS contract negotiation – contract approval	Access agreement	Yes
Andean Community	Access application – ABS contract negotiation – contract approval	Access agreement	Yes
African Union Model Law	Access application – ABS contract – approval	Access agreement	No

Contracts have also become the starting point to specify ABS conditions which are enshrined in national legislation. As we have seen, most academic literature on ABS also takes contracts as the tool and instrument to operationalize ABS regulations and principles. Hence, the discussion turns to their potential to satisfy the moral and economic interests of Providers and Users.

But some optimism arises through the Nagoya Protocol. It suggests that a global multilateral benefit-sharing regime may need to be considered in cases of transboundary genetic resources or for which PIC is not possible.[25] Discussion remains as to the exact meaning of "transboundary" or cases where PIC may not be possible, but the opening is there.[26]

Transboundary means somewhere *outside* national jurisdictions, beyond national frontiers. One immediate interpretation is the case of common resources shared by various countries or cases of *ex situ* collections of genetic resources, outside national borders. However, cases where PIC may not be possible could include resources from areas such as the high seas and the deep sea bed or Antarctica – which are governed under special legal and management regimes.[27] In such circumstances, PIC may be irrelevant and MAT never possible (in terms of fairness and equity) in the context of access to and utilization of natural information![28]

Interestingly, Article 11 of the Protocol also addresses a *de facto* very common occurrence: genetic resources and TK that are shared *in situ* by two or more countries.[29] The Protocol presents a common occurrence and reality as if it is an exceptional situation for the purpose of ABS. In these circumstances, countries must endeavor to cooperate to implement the Protocol. It makes sense that Article 11 is read in conjunction with Article 10 of the Protocol.[30]

Indicative of the potential implications of Article 10 of the Nagoya Protocol, the CBD Secretariat convened the Online Discussion solely devoted to Article 10 from March to May 2013.[31]

An intervention by Mr. D'Alessandro of the Federal Office for the Environment and Nature of Switzerland (FOEN) during the Online Discussion minces no words:

> In my understanding, in the context of the Nagoya Protocol, we are however neither talking about species, subspecies or any other taxonomic entity, nor are we talking about information. In the context of the Nagoya Protocol we are talking about genetic resources. And a genetic resource is defined as genetic material of actual or potential value according to Article 2 of the Convention. It is hard to imagine how a genetic resource as material can occur in two or more countries at the same time. I believe that in principle it should always be possible to determine the source of a specific material, and therefore the bilateral approach is probably applicable in most cases, no matter whether material with similar properties is found in different countries.
>
> (Comment #4953)

By adhering to the mistake of understanding genetic resources as a tangible, no transboundary situation exists. Nothing can be in two places at the same time. If the CBD were not a framework convention, the Online Discussion could have ended at that moment. Complementing the same line of thought as Mr. D'Alessandro, the Swiss Academy of Science has long recommended that simple procedures be created for ABS by Contracting Parties to the CBD: "stimulating initiatives by the Contracting Parties to create supportive instruments such as framework contracts and certification schemes" (Biber-Klemm *et al.* 2010: p. 23).

Ironically, the easy access sought can be best achieved not under the bilateral contracts advocated but under the alternative regime of bounded openness, where natural information can flow for both non-commercial and commercial research, impeded only by existing administrative permitting rules. The global *ex post* obligation to share a minimum standard of monetary benefits occurs once a patented biotechnology is commercialized (see Chapter 5 for further details).

To understand the advantage of a fixed royalty that reflects an economic rent, one needs only to review from known cases where access has been granted to genetic material (Table 2.4). In a context where genetic resources are understood as natural

Table 2.4 Monetary benefits negotiated in ABS contracts and agreements

Contract/agreement	Monetary benefit
INBIO	In the two cases where commercial products have been obtained from bioprospecting, the agreements between the user and INBIO involves payment of 3% of total net sales of the commercial product (Personal visit to INBIO and interview with researchers, 2011).
ICBG – Peru Project	Various agreements determine binding obligations between Searle, Washington University, National Natural History Museum, University Cayetano Heredia and Aguaruna indigenous peoples. Undisclosed royalty rate will be paid by the corporate partner (Searle) to the universities (25%–25%), national museum (25%) and communities (25%), when and if a commercial product is generated and is based on relevant research and TK. The royalties will be shared equally among universities and Aguaruna communities (Rosenthal 1999). What is interesting is that the percentage split can be disclosed but not what is being split! The percentages lead to confusion with the royalty, perhaps purposefully reiterating the point of GRAIN.
Bioamazonia (Brazil) and Novartis	In 2000, Novartis agreed to pay Bioamazonia US$4 million over a period of three years for access to Brazilian *in situ* Amazonian genetic resources, plus a 1% royalty for any commercial product developed (Quezada 2007).
Extracta Natural Molecules (Brazil)	Since 1998 bioprospecting for useful molecules from Brazilian biodiversity. It pays 2.5% royalties from sales of commercial products to individual property owners (Quezada 2007).
Institute of Biodiversity Conservation (IBC – Ethiopia), Ethiopian Agricultural Research Organization (EARO) and Health and Performance Food International (HPFI) (the Netherlands) – Teff Agreements	Four options: one-time (undisclosed) payment to IBC (based on HPFI net incomes in 2007-2008-2009); annual royalty payments to IBC (30% of net profits resulting from sales of basic and certified teff seed); or annual license fees to IBC based on number of hectares of teff seed grown (by company and licensees); annual contributions to fund (5% of HPFI net profits – never less than €20,000) (Andersen and Winge 2012). With profits as the base of calculations, the possibility exists for a 30% share of zero.

Note: Some of these agreements are no longer in place. The case of Teff is on the limits of ABS: it seems to involve the commercialization of a commodity – even though it may have been improved through research and breeding.

information, most benefits are being calculated based on different forms of value added (i.e. taxonomy, initial screening, breeding, and so on). However, there is no compensation for the opportunity costs of preserving the natural information, particularly in *in situ* conditions. In INBIO and the International Cooperative Group (ICBG) projects, and more recent experiences in Africa, Australia and Asia, the management and institutional arrangements regarding ABS are defined in contracts.

Several fallacies surface. The contracts reflect not just the actual text of the CBD but also a conflation of sovereignty over genetic resources with the right to negotiate bilaterally. In other words, the fallacy of begging the question (*petitio principii*) surfaces: bilateral contracts are an expression of sovereignty and because nations are sovereign, the contracts must be bilateral. The fallacy is compounded by the appeal to authority (*ad verecundiam*) of the IUCN Explanatory Guides and other influential texts and guidance from experts. No one seems to entertain the fact that a multilateral regime is also an expression of sovereignty.

Mutually agreed terms invariably include contractual clauses on confidentiality which are beyond examination much less debate. Hence, confidentiality makes a mockery of the criteria of fairness and equity in the sharing of benefits, especially monetary benefits. How can one comment on fairness and equity when key benefits are undisclosed? Advocates of bilateralism commit the fallacy of special pleading when they accept the norm and principles of fairness and equity in the ratification of the CBD, except when they apply to an ABS contract or MTA. With rare exceptions, bioprospecting projects, the classic form of ABS projects, do not disclose royalty rates, which are presumably very low (Markandya and Nunes 2012: p. 11).

Existing ABS national legislations and regulations in the Andean Community, Costa Rica, Panama and Peru, allow Users and Providers to invoke confidentiality in ABS contracts. If a specific ABS framework does not expressly invoke confidentiality, administrative, common or civil law rules also enable confidentiality as supreme over transparency.

In the US, a non-Party to the CBD and the Nagoya Protocol, the Federal Technology Transfer Act of 1986 protects against disclosure of certain types of information, such as royalty rates in collaborative agreements including bioprospecting projects between Federal agencies and private entities. In other countries if not by law, bureaucracies make it almost impossible to readily access detailed information about contractual provisions in ABS contracts and agreements regarding monetary benefit-sharing. Silence prevails in the name of competitive advantages and strategic or commercial interests of Users. Accessing specific clauses and figures in ABS contracts has always proven extremely difficult. One can safely speculate that secrecy under "confidential or sensitive business or commercial information" will become the standard in the vaunted model contracts.

The economic essence of standardization, except for the royalty rate, is to take Provider countries to the economic ideal of "perfect competition", where rents are completely eliminated (Vogel 2007b; West 2012). The Nagoya Protocol is an enabler of confidentiality and perfect competition. In the context of regulating checkpoints and monitoring of genetic resources, Article 17 determines that information provided to the designated checkpoint (on PIC, MAT, origin, etc.) will be without prejudice of the protection of confidential information. Though not specified, one can have no doubt whatsoever that the commercial and monetary benefit-sharing details will be withheld from the public.[32]

The rationale for secrecy may relate to the type of genetic resources being accessed and used, the location and even the type of research to be undertaken. The economic argument is that such information is useful to competitors who have not borne the associated costs and may take unfair advantages.

While monetary benefits are hardly ever made public, non-monetary benefits are trumpeted, apparently for public relations. The various "ABS case studies" produced since 1992 are based on an underlying assumption that they are good examples of bioprospecting projects or agreements. They invariably highlight non-monetary benefits but hardly ever mention royalty rates and occasionally indicate other forms of compensation, mostly upfront payments and milestone payments.[33]

There are various examples which evidence such selective silence in information. A few may be illustrative. The study by Jorge Cabrera (2009) on INBIO and bioprospecting in Costa Rica includes tables with data on partners involved, resources accessed, fields of research, time frames for projects, and so on. There is only a general list of monetary and non-monetary benefits with no specific reference to payment figures or royalties agreed. Other studies on INBIO exhibit the same lacunae.[34]

Kerry ten Kate's case study on Diversa's bioprospecting for *Thermus aquaticus* in Yellowstone Park in the US, is very detailed in regard to the upfront payments and milestone payments but royalty rates for future commercialization of successful products are undisclosed – due to legal and statutory constraints.[35]

Walter Lewis has analyzed the Peruvian ICBG project in detail, which involved two national institutions, a US corporation and a US university. Only a broad general reference is made to the distribution of negotiated royalty rate, 75 percent of which accrued to Peruvian parties. However, the actual rate on the successful commercial product is missing in this analysis.[36]

The CBD Secretariat has commissioned various similar case studies and all include similar blank spaces in regard to specificities when monetary benefits are at stake. The issue is not a trifling detail.

Where agreements have become public, the monetary benefits measured in royalty rates are mostly somewhere around 0.5 percent and 3.0 percent of sales of products derived from genetic resources. Often, Providers participate in a portion of these percentages. The arrangements on how often a royalty is paid are again difficult to obtain.

GRAIN, an active and very vocal NGO, expressed early in ABS debates concerns over the apparently intended confusion over royalty rates:

> Most benefit sharing agreements are obsessed with royalties. But often the real meaning of published royalty figures is unclear, sometimes they are even deliberately confusing. Typically, royalty figures shown in benefit sharing case studies quote a percentage of an undefined whole, or refer to a sub-percentage of an unknown fraction of product sales.
>
> (GRAIN 2000)

Upfront and/or milestone payments may also be included in contracts and agreements. Royalties for future commercially successful products vary but the common feature is that they are almost never publicized. Even the CBD Secretariat has at some point justified the need for confidentiality to ensure commercial interests of investors and entrepreneurs.[37]

The low royalty rates contrast starkly with the commercial value of genetic resources in the biotechnology sector which ranges from crop improvement and breeding, to the production of pharmaceuticals, cosmetics and industrial enzymes. Based on the ten Kate and Laird calculation for global annual markets for genetic resources of US$500–800 billion, the potential royalty incomes from contracts would only amount to an annual US$4 billion globally (at 0.5 percent). This seems hardly significant to incentivize global conservation.[38]

Although the variation in these figures is substantial, the magnitudes justify two crucial economic issues for ABS: the degree of substitution of genetic resources for other inputs in R&D and the costs of creating a multilateral system to capture benefits sufficiently significant to either incentivize habitat conservation or at least dampen land use conversion. Table 2.4 shows that royalty percentages as currently negotiated in ABS projects, contracts and agreements would probably achieve neither.

Secrecy and confidentiality in both ABS negotiations and the subsequent contracts make impossible any assessment of equity and fairness. The reasons may inhere to self-preservation. National authorities may be understandably reluctant to share information which would invite criticism of the meager benefits. Humans are territorial and also protect bureaucratic turfs.

The failure to secure significant royalties is not due to an asymmetry in negotiation skills between Users (e.g. communities) and Providers (e.g. public lands or *ex situ* collections). The asymmetry lies elsewhere: in the low marginal costs to collect physically and extract natural information and the high opportunity costs to maintain that information in habitats. Vogel uses the Australian case of effective average royalties of 0.3 percent as a natural experiment, given the presumably high negotiation skills of the Provider.[39] Australian legal scholar Peter Drahos concurs that the State/government is receiving "peanuts for biodiversity" (Drahos 2014).[40]

The economics is straightforward: with many countries harboring the same natural information (Table 2.5), the market price goes down to the marginal cost of collection in the country most willing to sell cheap. The nature of genetic resources as natural information is the reason and not, as many believe, because "poor people sell cheap" (Martinez-Alier 2005) and Providers tend to be poor. Australia serves as a control, being one of the few developed Providers. Hence, the paradox: genetic resources command a low value despite a trillion-dollar economy based upon them.

Multilateral systems of bilateral agreements also fail to generate monetary benefits. The *Standard Material Transfer Agreement* or SMTA under the ITPGRFA is, in practice, an adhesion contract, which means in layman's terms "take it or leave it". Monetary benefits derived from the commercial use of PGRFA included in the system are predefined and do not accrue specifically to countries harboring the same natural information.[41] At present, monetary benefits are used to support conservation projects in developing countries (of origin). A decade after its entry into force, no monetary benefits have been generated from the ITPGRFA Multilateral System. Donations and cooperation from industrialized countries are what makes it

Table 2.5 Useful products from natural information and their geographic distribution

Product	Use	Biological origin	Original distribution of species
Atropine	Anti-inflammatory, anti-cholinergic	*Atropa belladonna* (Solanaceae) (plant)	Native from Europe, northern Africa, western Asia (also naturalized in North America)
Chloroquine	Anti-malarial	*Cinchona ledgeriana* (plant)	Bolivia, Ecuador and Peru
Ephedrine	Anti-asthmatic	*Ephedra gerardiana* (plant)	Southwestern North America, southern Europe, northern Africa, and southwest and central Asia, northern China, and western South America
Hirudin	Anticoagulant	*Hirudo medicinalis* (animal – leeches)	Europe and Asia (particularly Kazakhstan and Uzbekistan)
Ryanodine	Pesticide	*Ryania speciosa* (plant)	Central and South America
Taxol	Anticancer	*Taxus brevifolia* (plant – tree)	North America (Pacific North West)
Vasotec	Anti-hypertensive	*Bothrops jararaca* (snake)	Brazil, Paraguay, Argentina
Vinblastine	Anticancer	*Catharansus roseus* (plant)	Madagascar (endemic)
Voltaren	Anti-inflammatory	*Salix* spp. (plant – tree)	Cold and temperate regions of the Northern Hemisphere
Zovirax	Antiviral	Unknown except that it comes from a Floridian marine sponge	
Mevacor	Cholesterol lowering agent	*Aspergillus terrestris* (fungi)	Non determined

possible to finance conservation projects – under the Benefit Sharing Fund which is fully operational and "simple" in concept.[42]

The Ad Hoc Advisory Committee on the Funding Strategy of the ITPGRFA recognizes "the flow of income to the Benefit Sharing Fund has stagnated, and that nothing indicated that this trend would reverse in the near future" and furthermore, that "no benefits deriving from the use of germplasm from the Multilateral System, had accrued to the Benefit Sharing Fund".[43] Efforts are under way to find ways in which this situation can be reversed.[44]

In sum, the problems of ABS contracts can be reduced to three interrelated points: negotiations will never achieve fairness or equity if the object of access is defined as genetic material, rather than information; asymmetries in awareness of the potential value of natural information between Users and Providers will persist, especially given the uncertain and increasing complexity in the nature of and pathways for R&D; developing countries may wrongly believe that their inability to secure significant benefits is due to asymmetric negotiating power rather than a competitive market over natural information that drives the price of it down to the marginal cost of accessing samples, which is negligible. Users will "forum shop" to secure the lowest price.

NOTES

1 For a summary of ABS issues and indigenous peoples' rights, see Tobin, B. (2009) "Setting Protection of TK to Rights – Placing Human Rights and Customary Law at the Heart of TK Governance," in Kamau, E.C., Winter, G. (eds.) *Genetic Resources, Traditional Knowledge and the Law. Solutions for Access and Benefit Sharing*, 101–118. London and Sterling, VA: Earthscan.

2 Decision 391 of the Andean Community is a regional regime on ABS which includes: an Access Contract between the State and the applicant; an Accessory Contract between the physical provider or institution of genetic resources and applicant; an Annex, which is an agreement between the applicant and an indigenous community for access to and use of TK; a Framework Access Agreement between an applicant and *ex situ* and/or research institutions and Administration; and Deposit Agreements for custodianship of genetic resources. Although not all six contractual forms need proceed simultaneously, at least three or four should so proceed. For a detailed review of Decision 391 see Rosell, M. (1997) "Access to Genetic Resources: A Critical Approach to Decision 391 'Common Regime on Access to Genetic Resources'." *RECIEL* 3(3): 274–283.

3 In 2002, a researcher in Venezuela, at the time a country of the Andean Community, demonstrated the complexities of Decision 391 and its immediate effect on only a handful of ABS agreements concluded in Venezuela. See Febres, M.E. (2002) *La Regulación de Acceso a los Recursos Genéticos en Venezuela*. Centro de Estudios del Desarrollo. Serie Mención Publicación, Caracas, Venezuela. Similar conclusions were reached in Bolivia. See also Zapata, B. (2004) *La Experiencia Boliviana en la Aplicación de la Decisión 391: Régimen Común sobre Acceso a los Recursos Genéticos*. Ministerio de Desarrollo Sostenible. La Paz, Bolivia. The Philippines and a few other countries all show very few bioprospecting agreements or contracts concluded under existing ABS legal frameworks. A group of researchers published a paper on the negative effects of ABS in an area of research overlooked by policymakers: biological pest control. See Cock M.J.W. *et al.* (2010) "Do New Access and Benefit Sharing Procedures Under the Convention on Biological Diversity Threaten the Future of Biological Control?" *BioControl* 55(2): 199–218.

4 Since its creation in 1989, INBIO has concluded more than 50 ABS related agreements with prestigious companies, universities and research institutions, such as Merck & Co., Bristol-Myers Squibb, Eli Lilly & Co., Diversa Corporation, Strathclyde University, University of Massachusetts, the Institute of Chemistry and Cell Biology, Harvard Medical School and the Swiss Tropical Institute. By the sheer *number* of agreements INBIO is the most successful model for bioprospecting and adding value to biodiversity. However, it has not produced commercially viable products to maintain its overall operations over time. See brief by Third World Network (2013) *Costa Rica's INBIO Nearing Collapse, Surrenders its Biodiversity Collections and Seeks Government Bailout*. April 2013. Available at www.twnside.org.sg/title2/biotk/2013/biotk130401.htm

5 Law 7788 (Biodiversity Law, 1998) and Decree 31514-MINAE (General Norms for Access to Elements, Genetic and Biochemical Resources of Biodiversity, 2003) apply almost exclusively to INBIO and its different research and bioprospecting activities in protected areas in Costa Rica. The national competent authority, which is known by the acronym CONAGEBIO, exists primarily to process INBIO or its partners' applications and monitor progress.

6 Brazil is also cited as an example of a success story, albeit a recent one. In a recent study on the implementation of the Provisional Measure 2.186-16/2001, 70 ABS agreements are referenced as having been concluded between 2001 and 2013, with 33 more agreements under review by the National Council for the Genetic Patrimony (CGEN). The focus is on access to the "genetic patrimony" for R&D. However, as is common practice, specific royalty rates or monetary benefits negotiated per agreement are not provided. See CNI (2014) *Study on the Impacts of the Adoption and Implementation of the Nagoya Protocol for Brazilian Industry*. May 2014. pp. 52–60.

7 Susette Biber-Klemm suggests the need for special, non-commercial research agreements to expedite basic and taxonomic research. See Biber-Klemm, S., Martinez, S.I., Jacob, A. and Jevtic, A. (2010) *Agreement on Access and Benefit Sharing for Non Commercial Research. Sector Specific Approach Containing Model Clauses*. SCNAT, Bern, Switzerland. Available at www.bfn.de/fileadmin/ABS/documents/6C33Ed01__2_.pdf

8 Article 8(a) of the Protocol (Special Considerations) establishes that:

> In the development and implementation of its access and benefit-sharing legislation or regulatory requirements, each Party shall:
> (a) Create conditions to promote and encourage research which contributes to the conservation and sustainable use of biological diversity, particularly in developing countries, including through simplified measures on access for non-commercial research purposes, taking into account the need to address a change of intent for such research.

9 Examples of model agreements and contracts are numerous. The World Intellectual Property Organization (WIPO) is developing a database on model and examples of ABS agreements (see www.wipo.int/tk/en/databases/contracts/). A recent publication by WIPO offers 20 examples of model contracts relating to a wide range of Providers and Users. See WIPO (2013) *Draft Intellectual Property Guidelines for Access to Genetic Resources and Equitable Sharing of the Benefits Arising from their Utilization*. Consultation Draft. February, 2013 (Appendix II) (available at www.cbd.int/financial/mainstream/wipo-guidelines.pdf). The *Standard Material Transfer Agreement* (SMTA) of the ITPGRFA is one such model (designed for a specific group of genetic resources). The Andean Community Decision 391 uses a standard model ABS contract (see Resolution 415, of July 22, 1996, available at http://intranet.comunidadandina.org/documentos/Gacetas/gace217.pdf). The Swiss Academy of Sciences has likewise proposed model non-commercial research ABS agreements, and so forth (see Biber-Klemm, S. *et al.* 2010). The Bonn Guidelines on ABS (Appendix I) offers Suggested Elements for Material Transfer Agreements (see www.cbd.int/doc/publications/cbd-bonn-gdls-en.pdf).

10 Early in the ABS debate, Gudrun Henne analyzed how PIC and MAT constitute a two-tier process: on one level is the State and on the other, Users and Providers of genetic resources, where PIC and MAT manifest themselves according to how countries regulate access procedures internally. See Henne, G. (1997) "'Mutually Agreed Terms' in the Convention on Biological Diversity: Requirements under Public International Law," in Mugabe, J., Barber, C., Henne, G., Glowka, L., La Viña, A. (eds.) *Access to Genetic Resources: Strategies for Benefit Sharing*, 25–53. Kenya: ACTS Press. See also Perrault, A. and Oliva, M.J. (2005) *Prior Informed Consent and Access and Benefit Sharing*. ICTSD/CIEL/IDDRI/IUCN/QUNO Dialogue on Disclosure Requirements: Incorporating the CBD Principles in the TRIPS Agreement on the Road to Hong Kong WTO

Public Symposium, Geneva, April 21, 2005. Available at www.ciel.org/Publications/ PIC_PerraultOliva_Apr05.pdf

11 The CBD Secretariat's official summary of the Online Discussion on Article 10 of the Nagoya Protocol includes numerous interventions which stress bilateral contracts as if it were the only approach to address and govern ABS and relations between User and Providers. For details see CBD Secretariat (2013) Synthesis of the Online Discussion on Article 10 of the Nagoya Protocol on Access and Benefit-sharing. UNEP/CBD/ ABSEM-A10/1/2. Available at www.cbd.int/doc/?meeting=ABSEM-A10-01

12 Even Free Trade Agreements (FTAs) such as that signed between the US and Peru include reference to contracts as the means to realize ABS. The Understanding on Biodiversity and Traditional Knowledge indicates that Parties "recognize that access to genetic resources or traditional knowledge, as well as the equitable sharing of benefits that may result from the use of those resources or that knowledge, can be adequately addressed through *contracts* that reflect *mutually agreed terms* between users and providers" (italics added). The preference for bilateralism transcends the CBD and influences many processes. For the complete text of the FTA between the US and Peru see https://ustr. gov/sites/default/files/uploads/agreements/fta/peru/asset_upload_file719_9535.pdf

13 See WIPO (2013: pp. 9–11).

14 These figures do not correspond strictly to ABS related activities but they are significant amounts. None of the existing ABS "success" stories (e.g. Taq/Yellowstone) have distributed benefits to Providers in a proportion which would reasonably be considered equitable in light of the profits generated.

15 Edward Hammond identifies 200 different patent applications claiming microbial genetic resources. See Hammond, E. (2014) *Patent Claims on Genetic Resources of Secret Origin. Disclosure Data from Recent International Patent Applications with Related Deposits Under the Budapest Treaty on the International Recognition of the Deposit of Microorganisms for the Purpose of Patent Disclosure.* Third World Network. February, 2014. p. 15 (available at www.twnside.org.sg/title2/series/bkr/pdf/bkr003.pdf). Oldham's data also documents ongoing bioprospecting and R&D with little or no regard for the CBD and ABS principles. See Oldham, P., Hall, S. and Forero, O. (2013) "Biological Diversity in the Patent System." *PLoS ONE* 8(11): 6 (available at www.plosone.org/article/ info%3Adoi%2F10.1371%2Fjournal.pone.0078737).

16 See Newman, D.J. and Cragg, G.M. (2013) "Natural Products as Sources of New Drugs over the 30 Years from 1981 to 2010." NIH. Public Access. Authors' Manuscript. July 24, 2013. Available at www.ncbi.nlm.nih.gov/pmc/articles/PMC3721181/pdf/ nihms356104.pdf

17 The concepts of "digital biopiracy" or "digital piracy" have entered the ABS discussion. According to the ETC Group, "Synthetic biology makes it possible to move genetic resources across borders as digital information. Agreements such as the Nagoya Protocol that govern 'material transfer' of genetic resources may become ineffective." See ETC Group. CBD COP 12 (2014) *Addressing Synthetic Biology.* Brief. Pyeongchang, Korea.

18 In reviewing ten capacity building programs on ABS over two decades, without exception all include a section on ABS contracts and/or training on negotiations of ABS agreements. Some examples are the SECO *ABS Management Tool: Best Practice and Standard Handbook for Implementing Genetic Resources Access and Benefit Sharing Activities* (2007); the ICIMOD, *Training for Trainers and Resources Manual. Access and Benefit Sharing from Genetic Resources and Associated Traditional Knowledge* (2009); the UNCTAD *Blended Learning Course on Biodiversity and Intellectual Property* (ASEAN) (2013); and for Spanish speakers, the UNESCO *Cátedra de Territorio y Medio Ambiente* training course.

19 See Echeverri, A. (2010) *Régimen Común de Acceso a los Recursos Genéticos: Biodiversidad y Separación de sus Componentes Intangibles y Tangibles.* Investigación CODI de

la Universidad de Antioquia. p. 157. Available at www.leyex.info/magazines/vol67n1496.pdf

20 See Tobin, B. and Taylor, E. (2009) *Across the Great Divide: A Case Study of Complementarity and Conflict between Customary Law and TK Protection Legislation in Peru*. Research Documents. Initiative for the Prevention of Biopiracy. SPDA. Year IV No. 11. May, 2009. Available at www.biopirateria.org/documentos/Serie%20Iniciativa%2011.pdf

21 See Maya ICBG Bioprospecting Controversy, available at http://en.wikipedia.org/wiki/Maya_ICBG_bioprospecting_controversy

22 See Centeno, J.C. (2009) *La Biopiratería en Venezuela*. Prensa Libre. p. 4. Available at www.rebelion.org/noticias/2009/5/85426.pdf

23 See Ploetz, C. (2005) "ProBenefit: Process-oriented Development for a Fair Benefit-sharing Model for the Use of Biological Resources in the Amazon Lowland of Ecuador," in Feit, U., Von den Driesch, M., Lobin, W. (eds.) *Access and Benefit-Sharing of Genetic Resources Ways and Means for Facilitating Biodiversity Research and Conservation while Safeguarding ABS Provisions*. Report of an International Workshop in Bonn, Germany, convened by the German Federal Agency for Nature Conservation. November 8–10, 2005. pp. 97–101. Available at www.bfn.de/fileadmin/MDB/documents/service/skript163.pdf

24 Many ABS model contracts and regulations include provisions for returning to the Provider or the national competent authority to negotiate monetary benefit-sharing when new circumstances arise in R&D. This is especially true for "non-commercial" initiatives which later reveal commercial/industrial potential.

25 Article 10 (Global Multilateral Benefit Sharing Mechanism) establishes that:

> Parties shall consider the need for and modalities of a global multilateral benefit-sharing mechanism to address the fair and equitable sharing of benefits derived from the utilization of genetic resources and traditional knowledge associated with genetic resources that occur in transboundary situations or for which it is not possible to grant or obtain prior informed consent. The benefits shared by users of genetic resources and traditional knowledge associated with genetic resources through this mechanism shall be used to support the conservation of biological diversity and the sustainable use of its components globally.

26 Article 10 of the Nagoya Protocol was included at the last minute in the negotiations in October 2010. Reasons for its sudden inclusion have not been voiced though it is known that the GMBSM was originally conceived by the African Group as "a new and innovative financial mechanism that would put into place a multilateral approach in parallel to the dominant bilateral modalities currently being negotiated." This idea may be traced to the 2007 Working Group on ABS No. 5. See the 2013 report by The Berne Declaration, Brot, ECOROPA, TEBTEBBA and TWN, *Nagoya Protocol on Access to Genetic Resources and the Fair and Equitable Sharing of Benefits Arising from their Utilization. Background and Analysis*. Penang, Malaysia.

27 The high seas and deep sea bed are managed under the United Nations Convention on the Law of the Seas (UNCLOS). Lyle Glowka was the first to elaborate the legal implications of ABS for the deep sea bed and high seas research, especially the extremophiles near hydrothermal vents. See Glowka, L. (2000) "Bioprospecting, Alien Species and Hydrothermal Vents: Three Emerging Legal Issues in the Conservation and Sustainable Use of Biodiversity." *Tulane Environmental Law Journal* 13: 329–360. Antarctica is governed by the Antarctic Treaty of 1959. In 1999, discussion began in regard to bioprospecting in the context of the Antarctic System. For a review of the regime applicable to bioprospecting in the Antarctic, see Lohan, D. and Johnston, S. (2005) *Bioprospecting in Antarctica*. Yokohama, Japan: UNU-IAS.

28 Prior informed consent is relevant only to the extent it coincides with the issuance of a certificate of origin which indicates a Provider and the species targeted. If unknown,

some description of the targeted population or taxon is identified which may lead at some point in the research process to a more precise taxonomic identification. Elimination of PIC is a critical element under the bounded openness regime proposed in Chapter 5.

29 Article 11 (Transboundary Cooperation) establishes that:

> 1. In instances where the same genetic resources are found *in situ* within the territory of more than one Party, those Parties shall endeavor to cooperate, as appropriate, with the involvement of indigenous and local communities concerned, where applicable, with a view to implementing this Protocol.
> 2. Where the same traditional knowledge associated with genetic resources is shared by one or more indigenous and local communities in several Parties, those Parties shall endeavor to cooperate, as appropriate, with the involvement of the indigenous and local communities concerned, with a view to implementing the objective of this Protocol.

30 The most convenient interpretation of Article 10 for maintaining bilateralism is that the high seas, deep sea bed or Antarctica are the only transboundary cases. Article 11, however, means shared resources despite the argument that "the same" cannot be in two jurisdictions simultaneously. The thesis of the book is that everything in the CBD and the Nagoya Protocol could be subject to revision in order to realize the objective of equity and fairness in benefit-sharing. Shared resources only make ontological sense if the object of access is not material.

31 See the official Synthesis of the Online Discussion at www.cbd.int/doc/?meeting= ABSEM-A10-01

32 Article 17(a)(iii) (Monitoring the utilization of genetic resources) establishes that:

> To support compliance, each Party shall take measures, as appropriate, to monitor and to enhance transparency about the utilization of genetic resources. Such measures shall include:
>
> (a) The designation of one or more checkpoints, as follows:
> > (i) Designated checkpoints would collect or receive, as appropriate, relevant information related to prior informed consent, to the source of the genetic resource, to the establishment of mutually agreed terms, and/or to the utilization of genetic resources, as appropriate;
> > (ii) Each Party shall, as appropriate and depending on the particular characteristics of a designated checkpoint, require users of genetic resources to provide the information specified in the above paragraph at a designated checkpoint. Each Party shall take appropriate, effective and proportionate measures to address situations of non-compliance;
> > (iii) Such information, including from internationally recognized certificates of compliance where they are available, will, without prejudice to the protection of confidential information, be provided to relevant national authorities, to the Party providing prior informed consent and to the Access and Benefit-sharing Clearing-House, as appropriate.

33 Various examples show gaps in information. Cabrera's study on INBIO and bioprospecting in Costa Rica includes data on partners involved, resources accessed, fields of research and time frames for projects but only a general list of monetary and non-monetary benefits. There is no specific reference to payment amounts (figures) or royalties agreed. Other studies on INBIO are no more illuminating; see Cabrera, J. (2009) "The Role of the National Biodiversity Institute in the Use of Biodiversity for Sustainable Development: Forming Bioprospecting Partnerships," in Kamau, E.C., Winter, G. (eds.) *Genetic Resources, Traditional Knowledge and the Law: Solutions for Benefit Sharing*, 244–269. London and Sterling, VA: Earthscan. ten Kate's study on bioprospecting *Thermus aquaticus* is very detailed with respect to upfront and milestone payments that Diversa pays Yellowstone National Park. However, royalty rates for future commercialization of successful products are undisclosed – due to legal and statutory constraints. See ten Kate, K., Touche, L. and Collins, A. (1998) *Benefit Sharing Case Study: Yellowstone*

Park and Diversa Corporation. Submission to the Executive Secretary of the Convention on Biological Diversity by the Royal Botanic Gardens, Kew, April 22, 1998 (available at www.cbd.int/financial/bensharing/unitedstates-yellowstonediversa.pdf). Lewis also analyzes in detail the Peruvian ICBG project, involving national institutions, a US corporation and a US university. Only a broad general reference is made to the distribution of negotiated royalty rate. Seventy-five percent of the royalty rate accrues to Peruvian parties. However, the actual rate is undisclosed. See Lewis, W., Lamas, G., Vaisberg, A., Corley, D.G. and Sarasara, C. (1999) "Peruvian Medicinal Plant Sources of New Pharmaceuticals (ICBG – Peru)," in Rosenthal, J. (ed.) "Drug Discovery, Economic Development and Conservation: The International Cooperative Biodiversity Groups". *Pharmaceutical Biology,* Volume 37, Supplement, Swets & Zeitlinger, the Netherlands. The CBD Secretariat has commissioned similar case studies that do not disclose the specific royalty rate.

34 See Cabrera, J. (2009) "The Role of the National Biodiversity Institute in the Use of Biodiversity for Sustainable Development: Forming Bioprospecting Partnerships," in Kamau, E.C., Winter, G. (eds.) *Genetic Resources, Traditional Knowledge and the Law: Solutions for Benefit Sharing,* 244–269. London and Sterling, VA: Earthscan.

35 See ten Kate, K., Touche, L. and Collins, A. (1988) *Benefit Sharing Case Study: Yellowstone Park and Diversa Corporation.* Submission to the Executive Secretary of the Convention on Biological Diversity by the Royal Botanic Gardens, Kew, April 22, 1998. Available at www.cbd.int/financial/bensharing/unitedstates-yellowstonediversa.pdf

36 See Lewis, W., Lamas, G, Vaisberg, A., Corley, D.G and Sarasara, C. (1999) "Peruvian Medicinal Plant Sources of New Pharmaceuticals (ICBG – Peru)," in Rosenthal, J. (ed.) "Drug Discovery, Economic Development and Conservation: The International Cooperative Biodiversity Groups". *Pharmaceutical Biology,* Volume 37, Supplement, Swets & Zeitlinger, the Netherlands.

37 In an interview with Intellectual Property Watch, regarding IP provisions and disclosure of certain information in patent applications, Mr. Braulio Dias, Executive Secretary of the CBD, indicates that "private companies are concerned about disclosure of information on the origin of materials." He further states that "It is a delicate issue for them because they do not want to lose competitiveness." See Chatterjee, P. (2012) "New Head of CBD: IPR Still Key to Nagoya Protocol on Access and Benefit Sharing." *Intellectual Property Watch Bulletin.* Available at www.ip-watch.org/2012/07/10/cbd-head-ipr-still-key-to-nagoya-protocol-on-access-and-benefit-sharing/

38 Similar studies have been attempted over the years. Though data and numbers also vary between them, they all concur that global markets in genetic resources are in the billions of dollars range and in one study, almost a trillion dollars. See ten Brink, P. (2009) "Rewarding Benefits through Payments and Markets," in *The Economics of Ecosystems and Biodiversity,* 34. TEEB.

39 Vogel, J.H. (2005) "Sovereignty as a Trojan Horse: How the Convention on Biological Diversity Morphs Biopiracy into Biofraud," in Hocking, B.A. (ed.) *Unfinished Constitutional Business? Rethinking Indigenous Self-Determination,* 228–247. Australia: Aboriginal Studies Press.

40 Peter Drahos examines the case of the Government of Queensland in Australia, and after reviewing the role of the Queensland Biotechnology Code of Ethics and the Biodiscovery Act, and a series of subsidies and tax breaks for private sector investments in bioprospecting through Federal and State compensation schemes for R&D, he concludes that what companies plan to pay for accessing the genetic patrimony of Australia is peanuts. See Drahos, P. (2014) *Intellectual Property, Indigenous People and their Knowledge,* 141–156. Cambridge: Cambridge University Press.

41 In summary, if a commercial product is obtained from access to and use of PGRFA from the Multilateral System, and is protected through a patent, the holder of the right will

pay to the Benefit Sharing Fund of the ITPGRFA 1.1 percent less 30 percent of the sales of the product (Article 6.7 and Annex 2, 1 of the SMTA). If the product is available without restrictions (e.g. protected by a breeder's right), the payment to the Fund will be voluntary (Article 6.7 and Annex 2, 2 of the SMTA). A third modality of payment involves 0.5 percent of the sales of any other products that are PGRFA belonging to the same crop (Article 6.11 and Annex 3, 1 of the SMTA).

42 The Multilateral System was developed in view of interdependence between countries in regard to PGRFA, critically important for food security. Case by case negotiation for PGRFA was considered a risky strategy given the need for continued flows and movements of resources among countries and institutions for breeding and conservation purposes. Benefits from a new plant variety would not cover the transaction costs involved, where developing countries are both Providers and Users of PGRFA for their agricultural systems. Inasmuch as many of the PGRFA in the ITPGRFA are not essential for food security, a more functional and pragmatic approach to addressing ABS for PGRFA was adopted. For an overview of interdependence as a justification for the ITPGRFA, see Andersen, R. (2008) *Governing Agrobiodiversity: Plant Genetics and Developing Countries*, 136–138. Aldershot: Ashgate.

43 See IT/ACFS-7 RES/13/Report, April 2013, para. 10.

44 For an analysis of this trend, see Correa, C. *ITPGRFA: Options to Promote the Wider Application of Article 6.11 of the SMTA and to Enhance Benefit-Sharing*. Legal Opinion. July 2013. The Berne Declaration, The Development Fund. Available at www.evb.ch/fileadmin/files/documents/Biodiversitaet/130731_Juristisches_Gutachten.pdf

Chapter 3

Sovereignty over genetic resources
The first twenty years of ABS

Insistence on fairness and equity in the utilization of genetic resources and associated traditional knowledge (TK) is manifest in the policy and legal frameworks of ABS. All include either direct or indirect references to "ensure", "guarantee" and "promote" the fair and equitable sharing of benefits arising from access to and use of genetic resources. Such emphasis on fairness and equity reflects an historic North–South political divide. A lack of trust clouds what could otherwise be a simple implementation of a technical solution for fair and equitable benefit-sharing. The neologism "biopiracy" is the most visible manifestation of the depth and breadth of political tensions and tacit distrust rooted in history.

An intensive flow of genetic resources, largely domesticated and wild plants, began in the fifteenth century with what the South calls plunder and the North, discovery. Contact may be the most neutral word. Potatoes moved directly from South America to Europe; coffee, indirectly, from northern Africa to Arabia, Europe and then the Americas; cotton, from Central Asia to Europe and Africa; maize, from Mesoamerica to Europe and beyond. Specific examples became highly transformative of the societies, economies and cultures of regions heretofore without contact.[1]

Evidence suggests that, at the time, equity and fairness considerations between kingdoms and States and "discovered" lands, in regard to exchanges and flows of biological resources and specimens were already generating tensions, under different manifestations. Wars for control of the spice trade can be traced back at least 4,000 years and extended to the modern ages,[2] and the emergence of botanical gardens in Europe in the sixteenth and seventeenth centuries were not solely for aesthetic and recreational purposes, but to enable reproduction of useful commercial and agricultural plants.[3]

The entry into force of the CBD in 1993 is a watershed in history precisely because of explicit language of fairness and equity to describe benefit-sharing resulting from the *utilization* of genetic resources. A new chapter of history was trumpeted.[4] The consequence of the category mistake of the CBD should disabuse the reader of any such enthusiasm. The sharing of monetary benefits remains unfair and inequitable for the economic reason elaborated in the previous chapters.

In brief, competition in artificial or natural information eliminates rents which result from the category mistake of defining genetic resources as "material" and then interpreting "sovereignty" very narrowly (Vogel 1997; Ruiz *et al.* 2010). For the purpose of utilization in monopoly intellectual property, genetic resources are in essence information, usually widely dispersed across jurisdictions and know no political borders, as emphasized by Oldham in his research on patents and biodiversity (see Chapter 1).

Measurement of equity and fairness in non-monetary benefit-sharing is inherently more difficult. How does one measure and compare the benefits of national scientific capacity building? Training programs? Multiple or co-authorship on publications? Transfer of outdated equipment? Development of infrastructure? Scholarships and internships? To answer whether they are fair and equitable under agreements and ABS projects and arrangements, one would have to monetize the benefits and add together. In contrast, royalty rates adopt the measuring rod of money, which enables an easier economic analysis. The fact that royalty rates are typically less than 1 percent and atypically more than 2 percent, lends themselves to a reasonable judgment of unfairness and inequity. Why would anyone expect a User to be fair with non-monetary benefits when demonstrably so unfair with the monetary portion of the equation? The contradiction also invites closer analysis.

Relatively few ABS contracts have been concluded during the first two decades of the CBD excluding the MTAs signed under the ITPGRFA.[5] Carrizosa undertook a comprehensive analysis during the first decade of the CBD (Carrizosa *et al.* 2004). Although such systematization has not been updated comprehensibly for the next ten years (2003–2012 and beyond), scattered references in the literature suggest the same trend. Where agreements or contracts are made public, many relate to national actors embarking on initial phases of projects, often for the purpose of taxonomic and non-commercial research.[6] In other cases, it is difficult to distinguish whether the agreements actually refer to access to genetic resources per se, or are grouped with agreements for collecting biological specimens or for natural product developments in the line of biotrade, commercial value-chain-type enterprises.

As mentioned in Chapter 2, various explanations have been offered for the meager compensation and limited sharing of benefits from the Providers' side. What is also interesting is how Users justify paying "peanuts". Invariably they deploy some version of the Marxian labor theory of value, albeit almost never identified as such: the huge amount of money spent on R&D justifies not paying anything for the natural resource. Value derives purely from labor. Nevertheless, one issue unites both Providers and Users: benefits will be eroded from high transaction costs of implementing ABS regimes with limited possibilities of tracking and monitoring who uses which resources (Ruiz 2003; Carrizosa *et al.* 2004; Ruiz and Lapeña 2007; De Jonge 2009; Kamau *et al.* 2010; Stoll 2013).

All the explanations for meager monetary benefits are not mutually exclusive with a more fundamental dynamic.[7] Vogel stresses that competition over natural information is the overarching reason and that all the other explanations are marginal and can be safely ignored. As mentioned in Chapter 2, Australia offers a prime example of an industrialized country with highly skilled contract negotiators but one still only able to receive meager monetary benefits from its ABS contracts or "peanuts" as one distinguished scholar has mentioned (Drahos 2014). Such ranking of causes casts the whole ABS debate in a very different light and leads to a less charitable assessment of CBD implementation and COPs' performances.[8]

Nevertheless, other ABS legal scholars perceive no overarching cause and see competition as one among many causes. To overcome the consequences of competition, they advocate a "common pools" or "global commons" in genetic resources as a means to move away from strict bilateralism and realize equitable benefit-sharing (Kamau and Winter 2013a; Halewood et al. 2013). The nomenclature hails from the political scientist Elinor Ostrom who won the Nobel Memorial Prize in Economics in 2009.[9]

Without any recognition of the informational nature of genetic resources, the "commons" approaches suggest that there are formal and informal frameworks which govern the ways in which certain types of genetic resources are accessed and managed. These would include certain group practices and "ways of doing business" which are accepted by the group in a decentralized and collective decision-making process. Examples include microbial collections, botanical gardens, marine research and traditional medicine practitioners. Such frameworks function with horizontal rules and principles which govern their actions and activities as collectives.

Although the drift away from bilateralism is welcome, the "commons" approach leaves uncorrected the category mistake of Article 2 of the CBD which defines genetic resources as "material". The immediate distortion that follows in its wake is competition among "commons" and we return to the same problem of elimination of rents, only somewhat abated.

To drift further away from bilateralism, one must grapple with invocations of sovereignty. The tacit interpretation of "sovereignty" restricted to bilateralism seems non-negotiable in the Rawlsian sense of attempting to secure socially *just* distribution of benefits. It lies at the core of the political position of the South.

An alternative interpretation of sovereignty is urgently required not just for Providers in the South but also for Users in the North, who have been stymied by ABS under the CBD obligations. History affords lessons that may shed light on the future of reform, and the CBD process is no exception.

As is the case with most international agreements, the CBD is the fruition of an arduous process whose history is happily forgotten. The antecedent literature is seldom referenced in the ABS discussions and publications. To understand the text of the CBD presented at the Earth Summit Rio '92 and the trajectory of its ABS provisions, one must first review the "history of the plant genetic resources [political] movement", which dates to the 1960s (Pistorious 1997) and includes the role of *ex situ* centers and botanic gardens as repositories of genetic resources.[10]

Prior to the CBD, the debate on the rights and management of useful genetic resources largely concerned seeds. The venue was the Food and Agriculture Organization (FAO) and the vehicle, the International Board for Plant Genetic Resources (IBPGR). Interest became heightened in the 1960s through the astounding success of Norman Borlaug, father of the "green revolution", which paved the way for International Agricultural Research Centers (IARCs).[11] Issues regarding the ownership of seeds, innovations and technologies soon surfaced. By the late 1970s, Pat Mooney, Jack Kloppenburg and Henk Hobbelink were exploring three overarching questions. Who should control plant genetic resources? Who should have legal rights? What if any is the role of IP?[12] Nuanced answers were emerging just as patents over biotechnological innovations were expanding, in both the US and Europe.

Kloppenburg, expressing his view on how developing countries should react, suggested that:

> Third World nations have little to gain from the quixotic pursuit of common heritage in plant genetic resources. But they have a great deal to gain through international acceptance of the principle that plant genetic resources constitute a form of national property. Establishment of this principle would provide the basis for an international framework through which Third World nations would be compensated for the appropriation and use of their plant genetic information.
>
> (Kloppenburg and Kleinman 1988: p. 194)[13]

Meanwhile industry was becoming ever more assertive in its desire to strengthen the rights of innovators through IP and expand their reach globally. Three game-changing events occurred in the 1980s in favor of the Agro-, Biotech- and Pharma- sectors:

1. The US Supreme Court ruled that a modified life form could be patented in the 1980 Diamond v. Chakrabarty decision.[14]
2. Intellectual property entered the Uruguay Round of the General Agreement on Tariffs and Trade (GATT) in 1986, upon specific request of the US, heavily lobbied by its pharmaceutical industry.[15]
3. The Bayh-Dole Act of 1980 allowed private investment in public institutions, including universities, particularly in the field of biotechnology.[16]

The drive toward enclosure of artificial information in the US was in full swing. Though slightly more conservative in its approach, Europe followed suit and started to make its IP laws more flexible and supportive of patents over biotechnologies.[17]

The first legal and policy international instruments addressing issues of rights over genetic resources were developed under the aegis of the FAO. Its FAO International Undertaking on Plant Genetic Resources (1983) was the first of a series of non-binding[18] legal instruments reflecting issues of sovereignty, rights of plant genetic resources, farmers' rights, intellectual property (plant breeders' rights), funding for conservation, participation in benefits derived from access to and use of plant genetic resources, among others.

The International Undertaking conceptualized plant genetic resources as "the common heritage of mankind", which was immediately contested by the developing countries of the South.[19] Cries of piracy of intellectual property rights by the North, largely over pharmaceutical patents, were met with similar outrage from the South over biopiracy. As McManis writes, both sides misused the term "piracy", as the South had, as of that time, not established the respective property rights over genetic resources of pharmaceuticals (McManis 2004). The rights of national communities were to be fundamentally reshaped by a series of parallel processes, including pro-patent court decisions and the GATT negotiations.[20]

The complaint over the appropriation of plant genetic resources found resonance in the complaint over land conversion for a multitude of development goals. The 1980s and 1990s evidenced the deleterious results of large-scale projects in many megadiverse countries of the South, including timber operations, hydroelectric dams and, mostly, the expansion of the agricultural frontier through internal colonization. Around the globe, mass extinction was afoot. Alarms sounded in scientific and academic circles and translated into a global political concern. The World Conservation Union (IUCN), the World Wide Fund for Nature (WWF), and other NGOs became authoritative voices about the rapid erosion and loss of biodiversity worldwide. In parallel, the United Nations, which had led the international environmental agenda through the Stockholm Conference on Human Environment (1972), commissioned a global environmental assessment, the Brundtland Report also known as *Our Common Future*, and undertook organization of the 1992 Rio Conference on the Environment and Development. The conference would launch the development of an internationally binding biodiversity conservation treaty[21] (Table 3.1).

Worried that Northern concerns for conservation would limit development in the South, the notion of sovereignty of countries in regard to their right to exploit and use their natural resources was rapidly integrated into the biodiversity discourse leading to the Rio '92 process. However, in a world with ever increasing interconnectivity and interdependence, the meaning and implications of sovereignty must be re-examined. Any notion of complete autonomy and unimpeded discretion duly invites criticism. Instead, sovereignty with respect to natural resources and the environment must be contextualized and its expression nuanced. The majority of multilateral environmental agreements include principles claiming the right of States to decide how to use natural resources. Laws and regulations from this principle are then developed into procedures for diverse natural resources (e.g. forests, soils, minerals, fisheries, hydrocarbons, and so on). However, compliance with these agreements may require relinquishing sovereignty in order to prevent transboundary pollution and degradation of ecosystems, facilitate compensation for the use of environmental services, cooperation in remediation strategies and co-management of transboundary water courses and so on. In a globalized world, sovereignty exhibits relativism.

In terms of sovereignty as seen from abroad, countries explicitly submit themselves to an international jurisdiction and obligations upon ratifying international agreements, which limit absolute rights.

Table 3.1 Pre-CBD and CBD milestones toward ABS

Year	Milestone	Reference
1972	The United Nations Conference on the Human Environment (non-binding)	International policy and legal principles pertaining to the environment and natural resources established
1980	World Conservation Strategy (IUCN, FAO, UNESCO, WWF, UNEP) (non-binding)	General principles for nature preservation
1982	The United Nations Working Group on Indigenous Populations	Calls for the preservation of nature and natural resources from the perspective of indigenous peoples' lands and territories
1982	The World Charter for Nature (non-binding)	Basic international principles for environmental protection
1983	*Our Common Future* – Brundtland Report	Call for measures to conserve and preserve the world's natural resources – concept of "sustainability" highlighted
1984–1986	IUCN Environmental Law Center (with collaboration of the IUCN/WWF Plant Advisory Group)	Draft articles on a biodiversity treaty
1987	UNEP Governing Council resolution 14/26	Recognizes the need for an international agreement on biodiversity conservation
1988–1991	Ad Hoc Working Group on Experts led by UNEP	Discussions on possible framework convention
1991	UNEP creates the Intergovernmental Negotiating Committee for a Convention	Formal negotiations for a biodiversity convention begin
1992	United Nations Conference on Environment and Development (UNCED)	CBD is adopted in May 1992
1993	CBD enters into force	
2000	Cartagena Protocol on Biosafety	International rules for transboundary movement of GMOs adopted
2001	ITPGRFA	International rules on ABS for plant genetic resources for food and agriculture
2004	Bonn Guidelines on ABS adopted (non-binding)	Recognition that international measures (including User measures) are required to realize ABS
2010	Kuala Lumpur/Nagoya Complementary Protocol to the Cartagena Protocol	Compensation, liability and redress regarding GMOs
2010	Nagoya Protocol on ABS	International rules on ABS (including mandatory User measures)

Complementary to this external type of sovereignty, States also rely on an internal sovereignty. Such sovereignty is recognized in various international instruments which grant States the power to decide how and under what conditions to organize themselves and determine how to assign rights to natural resources and how to define economic policy. For example, in the sphere of human rights or demarcation of borders, the international order has an important impact on what can and ought to be done within the country. In other words, the notion of sovereignty is highly circumscribed, notwithstanding the powerful rhetoric that may be trumpeted to arouse political support, especially during elections. History is again, informative.

Sovereignty can be traced to the Middle Ages and therefore predates not just industry and agriculture but also the Enlightenment (Box 3.1). Traditional interpretations have become shockingly anachronistic in an era of post-industrial globalization. The simple message of interdependence has penetrated the public sphere. For example, *New York Times* columnist Thomas Friedman reflects on how the world has not only become "flat" through globalization and technology in particular (Friedman 2007: p. 60), but also networked to the point that a small crisis in any given country has ramifications in all the rest (e.g. the financial crises of the Asiatic countries in the 1990s, the 2008 collapse on Wall Street). Nowhere was this truer than in the energy dependency of the majority of countries in the last half of the twentieth century: industry depended on a handful of producers and exporters, namely, the Organization of the Petroleum Exporting Countries (OPEC). Science and technology have also advanced in a way that concentrates expertise in a few countries, thereby limiting external sovereignty.[22] Hence, claims of sovereignty by countries also dependent on foreign R&D lose much of their meaning. So, one may challenge the argument that "sovereignty" would be abrogated through a multilateral system of ABS.

Box 3.1 The evolving meaning of sovereignty

"Sovereignty" emerged with the formation and consolidation of nation-states during the Middle Ages, as part of the struggle of royals against the Holy Empire, Papacy, and feudal lords. Authors such as Jean Bodin, Thomas Hobbes, Antonio de Padua, William of Occam, Niccolo Machiavelli and Jean-Jacques Rousseau began to systematically study the essential characteristics of the power structures, laws and institutions that shape the State. What is known as "sovereign power" is perpetual, inalienable and the source of law. With the French Revolution, the idea of sovereignty residing in the nation or people began to settle. The public power represented by the State was now seen as emanating from the nation or people. In essence, sovereignty is its own component and inherent in power and therefore, in the State.

After the formation of nation-states in Europe, the idea of sovereignty served to nourish independence and self-determination movements in the former colonies. Sovereignty has also been influential in the development of international law by putting States on an equal footing and thereby establishing the bases for the configuration of an international legal order, expressed in instruments such as treaties, conventions and soft law, starting from the principle of *pacta sunt servanda* (promises must be kept). For practical purposes, sovereignty can be understood as an attribute of power exercised by a State that in turn is comprised of a territory and a people.

Sovereignty legitimizes the State to decide, by means of its constitution, how and under what conditions to assume international obligations and limit its conduct in certain spheres (e.g. commerce, human rights, border control, the environment and so on) vis-à-vis other States likewise obligated under an agreement or international treaty. A State finds itself obligated by international legal order and rules of conduct to moderate its behavior.

Sovereignty is routinely invoked in international agreements, such as in the form: "that any State has the sovereign right to ban the entry or disposal of foreign hazardous wastes and other wastes in its territory" (Basel Convention on the Control of Transboundary Movement of Hazardous Wastes and their Disposal 1989) or, conforming to the Charter of the United Nations, "the sovereign right to exploit their own resources pursuant to their own environmental and developmental policies, and the responsibility to ensure that activities within their jurisdiction or control do not cause damage to the environment of other States or of areas beyond the limits of national jurisdiction" (UN Framework Convention on Climate Change, 1992). Similarly, the Vienna Convention on the Law of Treaties (1969) commits to take into consideration "the principles of international law embodied in the Charter of the United Nations, such as the principles of the equal rights and self-determination of peoples, of the sovereign equality and independence of all States." As a final example, the Charter of the United Nations (1945) determines that the organization "is based on the principle of the sovereign equality of all its Members." *Seemingly paradoxical, the adoption of an international instrument implies a sovereign act of a State which has willingly accepted obligations.*

The CBD is oblivious to such nuances. The Preamble establishes "that States have sovereign rights over their own biological resources"[23] and in Article 15(1) that "[R]ecognizing the sovereign rights of States over their natural resources, the authority to determine access to genetic resources rests with the national governments and is subject to national legislation."

The misplaced sanctity of sovereignty and the definition of "genetic resources" as "material" guarantee an ABS and contractual system where there is no fairness and equity in benefit-sharing. Although countries may wrongly insist that only bilateralism is an expression of sovereignty, genetic resources do not oblige to physical boundaries. What is found in one country may be found in a neighboring country or even worldwide.

Nevertheless, the simple reality of diffusion of genetic resources across jurisdictions and subsequent competition among Providers is finally resonating in the more popular accounts of ABS. For example, Claudio Chiarolla acknowledges that:

[I]n transboundary situations some genetic resources can be shared between several countries which become de facto competitors. ... Historically ... limited or no value from the utilization of genetic resources has accrued to provider countries, which are left with no additional incentives to conserve biodiversity *in situ.*

(Chiarolla *et al.* 2013: p. 3)

Other authors also recognize that the issue of transboundary and/or shared genetic resources is a problem in terms of ensuring equity and fairness in benefit-sharing but most overlook the impact of competition on the royalty rate. Hence, they fail to explore the economics even when they teeter on the obvious implications for a multilateral system that fixes the royalty rate and provides a scheme for distribution of the revenues.

The 1996 Andean Community Decision 391 on ABS is a good example. It acknowledged the obvious: instances where the same genetic resource can be accessed in various Member States[24] and calls for countries to cooperate "and take into account the interests of countries which share the same genetic resources." Twenty years on, no collaboration between countries has yet occurred or is contemplated. The Nagoya Protocol similarly recognizes the issue of transboundary resources and those shared by more than one country, and suggests *ad hoc* solutions at the international level. Although unsatisfactory, Articles 10 and 11 of the Nagoya Protocol are, at least, an opening.

Claiming sovereign rights over transboundary or shared resources becomes meaningless when the economic implication is that competition over perfect substitutes for biotechnological research will eliminate all rents.

Paradoxically and counterintuitively biopiracy is also facilitated under the bilateral approach as no one may know which country of origin was the victim.

Such concerns are heightened to the extent that smaller and smaller biological samples are needed to access natural information. Many physical samples are transported undetected.[25] Others are natural extracts which find themselves in a non-ratified Party where no ABS obligation is binding. The US is most notable and problematic in this regard. However, as discussed in Chapter 5, the COP can address non-Parties through a system of bounded openness.

Inasmuch as sovereignty as commonly invoked renders impossible the third objective of the CBD, the COP needs to ask how sovereignty can be interpreted to assure the fair and equitable sharing of benefits. The answer begins by contrasting sovereign rights for tangibles versus intangibles. Most natural resources are tangible, including timber, minerals, fisheries and hydrocarbons, so sovereignty means bilateral accords in the various processes of extraction (i.e. logging, mining and fishing and so on). The advantages of asserting bilateralism as an expression of sovereignty to forests, fisheries and land is indisputable and understandably non-negotiable, especially for former colonies. Controls and enforcement are relatively straightforward and attainable.

In that historic frame of mind, the CBD was a facile attempt to broaden sovereignty and include genetic resources within its reach. Studiously ignored was the fact that the genetic resources are unlike the aforementioned natural resources. As information, the economics of genetic resources is not only distinct from but opposite to that of timber, minerals and other tangibles.[26]

A different interpretation of "sovereignty" in the light of both the nature of genetic resources as information and the economics of information should mean that countries have the power to decide to participate in a global multilateral regime which enables the capture of an economic rent. This can help offset the high opportunity costs of conservation.[27] The rent need not compete with the value in changes to land use. It is merely an incentive to counter political pressures to change land use as most values of habitats cannot be monetized.[28] Incentives are also realpolitik.

The regime would impose rules that render an effective "oligopoly in natural information". Vogel preferred to call it a "biodiversity cartel" in 1993, without any apology.[29] Although the term was meant to provoke, the desired discussion did not happen. Nevertheless, the idea has pierced the public sphere, and the less charged term from the Nagoya Protocol, "global multilateral regime on ABS", is more politically palatable.

As an aside, one notes that the Group of Like-Minded Megadiverse Countries (GLMMC) formed in 2001 was once considered a budding cartel and perhaps a precursor to support the GMBSM. Curiously, members of the Group never pursued the declared objective of "harmonization" of benefits, largely due to a steadfast adherence to the misinterpretation of sovereignty as solely implying bilateralism.[30]

A similar critique can be made of regional approaches to ABS in the Andean Community, the draft Central America Protocol on Access to Genetic and Biochemical Resources and Traditional Knowledge (1998),[31] the African Union Model Law for the Protection of the Rights of Local Communities, Farmers and Breeders, and for the Regulation of Access to Biological Resources (2000). All these approaches intuited the need for a common framework based on the reality that ecosystems are shared and genetic resources transboundary in essence.

NOTES

1 Hobhouse, H. (1999) *Seeds of Change: Six Plants that Transformed Mankind*, 4th edn, London: Papermac.
2 With the invention of tacking, sailing against the wind became easier and more efficient. "The first country that successfully circumnavigated Africa was Portugal, and in 1497 four vessels under the command of Vasco da Gama rounded the Cape of Good Hope, eventually sailing across the Indian Ocean to Calicut, India. This success marked the beginning of the Portuguese Empire. Spanish, English and Dutch expeditions soon followed, and the growing competition sparked bloody conflicts over control of the spices trade. As the middle class grew during the Renaissance, the popularity of spices rose. Wars over the Indonesian Spice Islands broke out between expanding European nations and continued for about 200 years, between the fifteenth and seventeenth centuries." See *The Silk Road. History of the Spice Trade*, available at www.silkroadspices.ca/history-of-spice-trade
3 "Gardens such as the Royal Botanic Gardens, Kew and the Real Jardín Botánico de Madrid were set up to try and cultivate new species that were being brought back from

expeditions to the tropics." See *The History of Botanic Gardens*, available at www.bgci.org/resources/history/

4   Perhaps italics are not enough emphasis: parties to the CBD have embarked on policy and regulatory measures centered on *access* as the trigger for benefit-sharing. Bounded openness suggests that the trigger for benefit-sharing should be utilization. Access centered on the initial physical access to samples containing genetic resources has given way to highly restrictive frameworks which run counter to the spirit and wording of the CBD. A recent study by Natural Justice on the EU Proposal for the Implementation of the Nagoya Protocol concludes that the regulatory burden of ABS frameworks should not be placed on physical access to genetic resources but rather on the moment of utilization – in bioprospecting, R&D. Bounded openness facilitates access even further by imposing disclosure at the moment of intellectual property registration, thereby also allowing easy monitoring of utilization. See Natural Justice, The Berne Declaration (2013) *Access or Utilisation – What Triggers User Obligations? A Comment on the Draft Proposal of the European Commission on the Implementation of the Nagoya Protocol on Access and Benefit Sharing.* Available at http://naturaljustice.org/wp-content/uploads/pdf/Submission-EU-ABS-Regulation.pdf

5   Under the ITPGRFA ABS Multilateral System, thousands of SMTAs – adhesion contracts – have been concluded, especially between the Consultative Group for International Agricultural Research (CGIAR) of *ex situ* genetic resources and Users from developing countries. One source indicates that for the period July 2007 to December 2010, 1,222,000 samples of accessions of PGRFA were transferred using the SMTA. Ninety-five percent were between CGIAR centers and developing countries. Nevertheless, no commercial/monetary benefits have flowed into the Benefit Sharing Fund of the Treaty since the Treaty entered into force a decade earlier. These SMTAs legitimize the facilitated flow of a specific set of PGRFA – established as part of a fixed list – which includes 35 species of PGRFA and 29 forage species. Led by the ITPGRFA Secretariat, efforts are under way to review alternative means of generating monetary benefits from access to and use of PGRFA. See Correa, C. *ITPGRFA: Options to Promote the Wider Application of Article 6.11 of the SMTA and to Enhance Benefit-Sharing.* Legal Opinion. July 2013. The Berne Declaration, The Development Fund. Available at www.evb.ch/fileadmin/files/documents/Biodiversitaet/130731_Juristisches_Gutachten.pdf

6   The situation in the Andean Community is illustrative. Overwhelmingly, agreements in Bolivia, Colombia, Ecuador and Peru cover national taxonomic research. For a detailed assessment of the implementation of Decision 391, see Ruiz, M. (2011) *Diseño de un Plan de Fortalecimiento de Capacidades Institucionales en el Tema de Acceso a los Recursos Genéticos Asociados a los Conocimientos Tradicionales.* Diagnóstico Regional y Anexos. Documento de trabajo. BIOCAN. Comunidad Andina. December 22, 2011 (available at http://biocan.comunidadandina.org/biocan/images/documentos/TallerARG/diagnostico_abs_documento_trabajo.doc). Costa Rica offers examples of dozens of contracts and agreements concluded between Users and INBIO, essentially the exclusive Provider for commercial and non-commercial use. The Costa Rican experience is unique in three aspects: first it provides access to natural information which is almost certainly shared by neighboring countries and perhaps even the entire neotropics; secondly, one suspects that Costa Rica does not obtain a rent from access but is being rewarded by value adding through taxonomy and initial R&D results – one does not know inasmuch as confidentiality is maintained over the royalties; thirdly, Costa Rica's ABS legislation is applied almost exclusively to regulate INBIO's activities. In the case of Brazil, the focus of the CGEN is access to the "genetic patrimony" for R&D. Royalties vary between 0.5 and 5 percent of net sales of products using genetic resources. But the same report also admits that in practice royalties are below 0.77 percent of net sales. See CNI (2014). *Study on the Impacts of the Adoption and Implementation of the Nagoya Protocol for Brazilian Industry.* May 2014. pp. 52–60. In informal discussions with a Brazilian

advisor to CNI, the advisor indicated that if the success of the Brazilian ABS regime is so successful, it is rather a paradox that modifications to Provisional Measure 2.186-16/2001 have been proposed by the government since at least 2010 (personal conversation at COP 12, October 15, 2014). The actual ABS contracts are not publicly available.

7 Roger Sedjo recognized that the issue of shared resources would become an issue in the development of policy and institutional structures to define ABS agreements, though he did not develop his argument in terms of the economics applied to natural information. See Sedjo, A.R. (1989) "Property Rights for Plants." *Resources* (Fall, 97): 3.

8 By 1991 Vogel had already coined the concept of "natural information" and emphasized diffusion over jurisdictions in an in-house publication: Vogel, J.H. (1992) *Privatization as a Conservation Policy: A Market Solution to the Mass Extinction Crisis,* 170. Melbourne, Australia: CIRCIT. This was to later be re-edited and published as Vogel, J.H. (1994) *Genes for Sale.* New York: Oxford University Press.

9 Elinor Ostrom (2009 Nobel Memorial Prize in Economics) "demonstrated that within communities, rules and institutions of non-market and not resulting from public planning can emerge from the bottom up to ensure a sustainable, shared management of resources, as well as one that is efficient from an economic point of view." See Felice, F. and Vatiero, M. "Elinor Ostrom and the Solution to the Tragedy of the Commons". *Il Sussidiario,* June 27, 2012. Available at www.aei.org/article/economics/elinor-ostrom-and-the-solution-to-the-tragedy-of-the-commons/

10 Lucile Brockway describes the specific role of Kew Botanic Gardens in the so-called "botanic chess game" of collecting, conserving and improving plants from distant British colonies for aesthetic, commercial and industrial interests. The analysis shows how botanic gardens are not only about scenery, beauty and recreation, but also about commercial and industrial potential. The collections of Kew, regardless of origin, are the Crown's property by law. See Brockway, L. (1979) "Science and Colonial Expansion: The Role of the British Royal Botanical Gardens." *Interdisciplinary Anthropology* 6(3): 449–465. Available at www.jstor.org/stable/643776

11 See Andersen, R. (2008) *Governing Agrobiodiversity. Plant Genetics and Developing Countries,* 87–91. Aldershot: Ashgate.

12 Together, Pat Mooney, Jack Kloppenburg and Henk Hobbelink produced some of the first critical and thought-provoking pieces on the politics, economics and law as applied to seeds and genetic resources. Thirty years after publication, their work is still relevant and much cited. Mooney focused on the privatization of genetic resources by agro-industry in the developed Northern countries and the role of biotechnology and modern breeding in this process. See Mooney, P.R. (1979) *Seeds of the Earth: A Public or Private Resources?* Ottawa: Canadian Council for International Cooperation and the International Coalition for Development Action. See also Mooney, P.R. (1983) *The Law of the Seed – Another Development and Plant Genetic Resources.* Dag Hammarskjöld Foundation. Available at www.dhf.uu.se/pdffiler/83_1-2.pdf. Kloppenburg followed the trajectory of plant breeding and focused on how "commoditization" has led to the development of modern biotechnological interests in plant breeding and an intensive use of intellectual property. See Kloppenburg, J. (1988) *First the Seed: The Political Economy of Plant Biotechnology.* Cambridge: Cambridge University Press. Hobbelink highlighted inequities and imbalances in rights and control over plant innovations, vastly concentrated in Northern industrialized countries and vested in corporate interests. See Hobbelink, H. (1991) *Biotechnology and the Future of World Agriculture.* London: Zed Books.

13 Kloppenburg would later change and suggest that genetic resources should be under State property or sovereignty. See Kloppenburg, J. (2005) *First the Seed: The Political Economy of Plant Biotechnology,* 2nd edn. Science and Technology in Society Series. University of Wisconsin.

14 The US Supreme Court ruled in favor of accepting a patent over a genetically modified bacteria. In its decision it states that "[T]his human-made, genetically engineered bacterium is capable of breaking down multiple components of crude oil. Because of this property, which is possessed by no naturally occurring bacteria, Chakrabarty's invention is believed to have significant value for the treatment of oil spills." See http://supreme. justia.com/cases/federal/us/447/303/case.html#F2

15 The history of the GATT since 1940 is described in Mindreau, M. (2005) *Del GATT a la OMC (1947–2005): La Economía Política Internacional del Sistema Multilateral de Comercio*. Universidad del Pacífico. Lima, Peru.

16 Bayh–Dole Act or Patent and Trademark Law Amendments Act (Pub. L. 96-517, December 12, 1980).

17 The Rote-Taube case (Bundesgerichtshof 1 IIC136, 1969), decided by the German Federal Supreme Court, opened the possibility for the protection of biologically derived innovations and inspired similar case law in the rest of Europe. In simple terms, the Court argued that there was no technical reason not to include biologically derived inventions within the scope of patent protection and its criteria.

18 Although non-binding, the International Undertaking adopted through FAO Resolution 9/83 and subsequent interpretative resolutions have become a strong driver of the genetic resources agenda and a milestone in defining certain issues in the negotiation of the CBD, including ABS, IP, technology transfer and TK. The Commission on Plant Genetic Resources (FAO Resolution 9/83) was at the time the main intergovernmental institutional setting within which implementation of the Undertaking and related policies were discussed.

19 The International Undertaking was premised on the "principle that plant genetic resources are a heritage of mankind and consequently should be available without restriction" (Article 1). Adherents would make the resources available for breeding and conservation "free of charge, on the basis of mutual exchange or mutually agreed terms" (Article 5). In 1991, FAO recognized that plant genetic resources were subject to the sovereign right of States (FAO Resolution 3/91), meaning that countries could exercise this sovereignty through regulatory measures and conditions.

20 Extension of patents or breeders' rights over biodiversity related innovations, in many cases using and modifying genetic resources and also using related TK in early phases of biological collecting and R&D processes, could be interpreted as disenfranchisement of countries and communities. In some cases it can be seen as a form of "misappropriation" or "biopiracy" of resources, regardless of the legal validity of patents and breeders' rights under existing IP rules and principles.

21 For a brief review of the history and background of the CBD, see Glowka, L., Burhenne-Guilmin, F. and Synge, H. (1994) *Guide to the Convention on Biological Diversity* Environmental Policy and Law Paper No. 30, Gland, Switzerland, pp. 1–7. For a more detailed analysis of the CBD, its background and initial developments, see McGraw, D. (2000) "The Story of the Biodiversity Convention: Origins, Characteristics and Implications for Implementation," in Le Prestre, P.G. (ed.) *The Convention on Biological Diversity and the Construction of a New Biological Order*, 9–42. Aldershot, UK: Ashgate.

22 Even though Brazil, China and India have industrialized, they still constitute a small group of countries compared to the US, Canada, Japan and the EU where biotechnology is highly concentrated.

23 The notion "own biological resources" does not specify a particular type of ownership right and is merely an abbreviated form to refer to resources that find themselves under the jurisdiction of the State, over which there may be converging rights by different stakeholders. See Glowka et al. (1994) as in note 21.

24 The Andean–Amazon regions are a geographical continuum which share ecosystems and biodiversity. Some of the better known and iconic cases of bioprospecting and

misappropriation involve resources shared by more than one country, e.g. maca, the poison dart frog, ayahuasca and quinoa.

25 One noteworthy experience is the National Commission for the Prevention of Biopiracy created in Peru through Law 28216 in 2004. Its role is to identify potential biopiracy cases as defined in the law and undertake legal or administrative actions nationally and internationally. The Commission has identified over 30 different species of Peruvian origin and hundreds of related patents, in some of which access was achieved with no PIC or MAT, thereby falling within the scope of the definition of "biopiracy". The Commission faces an almost insurmountable hurdle to determine how and when samples, seeds and specimens of these species left the country. See www.biopirateria.gob.pe/index2.htm

26 To better appreciate the difference between artificial and natural information, see Vogel, J.H. et al. (2011) "The Economics of Information, Studiously Ignored in the Nagoya Protocol on Access to Genetic Resources and Benefit Sharing." *Law, Environment and Development Journal* 7(1): 54–55. Available at www.lead-journal.org/content/11052.pdf

27 Oligopoly refers to sellers that reach agreement to not compete among one another, fix a price and produce a scheme to distribute rents. Although economic theory recommends against oligopolies, the recommendation is in the context of tangibles. For intangibles, the economic argument inverts.

28 This point is made in Vogel, J.H. (1997) "White Paper: The Successful Use of Economic Instruments to Foster the Sustainable Use of Biodiversity: Six Cases from Latin America and the Caribbean." *Biopolicy Journal* 2 (Paper 5). Available at www.bioline.org.br/request?py97005

29 Joseph H. Vogel has long advocated for a cartel. Others like Zamudio, Angerer, Omar-Oduardo and Ruiz have also expressed support and in recent years undertaken studies toward this end. There is a long trajectory and logical sequence to the idea of the cartel which begins with the proposal for a "Gargantuan Database" and a fair and equitable distribution of benefits, based on the extension of habitat conserved of the accessed species. The Gargantuan Database was first proposed in Vogel, J.H. (1992) *Privatisation as a Conservation Policy* (Melbourne, Australia: CIRCIT) (thereafter turned into *Genes for Sale*). Various other works have followed. For a recent reflection on "cartelization", see Vogel J.H. et al. (2011) "The Economics of Information, Studiously Ignored in the Nagoya Protocol on Access to Genetic Resources and Benefit Sharing" *Law, Environment and Development Journal* 7(1): 58–59. Available at www.lead-journal.org/content/11052.pdf

30 The GLMMC was formed in 2002 in Mexico. Its current members are Bolivia, Brazil, China, Colombia, Costa Rica, Ecuador, India, Indonesia, Kenya, Malaysia, Mexico, Peru, the Philippines, South Africa and Venezuela. The *raison d'être* was to coordinate a common political approach to ABS and related issues such as TK protection in various forums. The group collectively possess over 70–80 percent of the world's *in situ* biodiversity.

31 The draft Protocol was originally developed in 1997 as part of the Central America Commission on the Environment and Sustainable Development formed by Belize, Costa Rica, El Salvador, Guatemala, Honduras, Nicaragua and Panama. The Preamble refers to *shared and common resources* in Central America. However, the Protocol has not been approved or implemented given political uncertainties in regard to its implications for individual countries.

Chapter 4

Resistance to correction

The core idea of "bounded openness" for ABS is not new and antecedents can be identified even before the presentation of the CBD at the Earth Summit in June 1992. Understanding why the obvious and fairly simple solution has not been pondered or welcomed demands reflection and a plausible explanation. This chapter examines possible reasons for the dismissal. On many occasions since 1994, various aspects of the idea of bounded openness have surfaced in parallel events to the COP as well as in many ABS meetings, workshops and forums around the world. The literature is also abundant and accessible. To my knowledge, not once has anyone voiced counter-argument other than the non-argument of *stare decisis* (stand by the decision) and pointing out political unviability and that "it's too late in the process." Amazingly, these responses prevail and persuade. Why?

Answers lie at the interface of the natural and social sciences, which also pre-date the CBD. In the 1980s, the notion of "path dependency" had fully penetrated the economic literature, illustrated with examples that were commonplace such as the QWERTY typewriter keyboard, the convention of driving on the left or right side of the road, VHS versus BETA models. All the choices made sense at the time, but have become anachronistic and costly to reverse. The simplicity of path dependency means that from a given starting point, say the negotiation of the CBD in Nairobi in 1992, accidental or random events, say the personalities of negotiators, can have a significant and enduring effect on the outcome (e.g. bilateralism and contracts as the basis for ABS). Reversal in the prevailing trend is difficult due to positive feedback mechanisms, increasing returns and self-reinforcement.[1] The importance of setting the efficient standards early on became apparent as a way to eliminate the transaction cost of a future correction.

Perhaps had policymakers been more conversant with the trends in economic theory, and emerging research and advancements in science and technology applied to genetic resources, they would have given greater thought to what path they were embarking on before they embraced bilateralism. Though most policy and legislation can respond to charged circumstances, the gap between advances in science and technology and ABS frameworks continues to widen.

Reversing the policy and regulatory architecture in ABS is like changing the design on the keyboard or expecting the British to drive on the right hand side of the road for those accustomed to the left. These changes incur tremendous costs and resistance. The criterion for reversal should be: do the future benefits of the change outweigh the costs? Inasmuch as the bilateralism through ABS contracts and agreements seems to have failed so visibly over time, one would respond resoundingly, yes![2] Yet no such response emanates from either the COPs or the ABS process. Again, why?

The answer may be certain hostility from policymakers and a broad set of actors[3] driving the CBD and ABS process, toward basic economic thinking and science in general. But resistance to technocracy is not a unique feature of the CBD process. Examples abound in many realms. The current favorites of Professor Paul Krugman, the 2009 Nobel Memorial Laureate in Economics, seem to be the pitiful fiscal stimulus during the Great Recession and the underwhelming initiatives on climate change.

Rather than delve into the causation, he accepts as a given that "policy makers just keep finding reasons not to do the right thing." The insight applies well to ABS decision makers and actors who have made non-negotiable any correction of a glaring error in the CBD process: the definition and understanding of "genetic resources" as material and sovereignty as synonymous with ownership and an obligation toward realizing PIC and negotiating MAT through contracts.[4]

Krugman identifies a process of "intellectual devolution" where acceptance of a logical alternative is not as straightforward as it would seem – especially in the realm of policymaking. The issue of ABS puts intellectual devolution in high relief as the discussion is not only detached from logic but also the reality of biotechnology and R&D, and the failures of ABS projects to deliver equitable and fair returns from multi-billion dollar industries in pharmaceutical research, health care, cosmetics, crop protection, plant breeding and seeds. Failure is embarrassingly obvious. Nevertheless policy, enabled through secrecy, moves in the opposite direction from the viable alternative.

The unwillingness of advocates for the current model of bilateral ABS to even seriously consider or discuss alternatives such as "bounded openness" (Box 4.1) may be due to the unpersuasive power of logic and evidence in the realm of policymaking. The physicist Lawrence Krauss laments that technical and scientifically grounded advice is routinely ignored and that until such advice is appropriately assessed and acted upon, humanity faces grave and even existential threats.[5]

How to "translate" and streamline economics and science in ABS policymaking is a recurrent theme that dates back to the early 1990s, when a group of ABS actors, lawyers included, began calling for more precise and understandable scientific input into ABS policy processes.[6] Unfortunately, the call has not been heeded, thus echoing Krauss's concern.

Box 4.1 The sequence of elements for a bounded openness based ABS regime

1. Classification and definition of genetic resources as natural information

    +

2. Recognition that a multilateral regime (a GMBSM) is an expression of sovereignty

    +

3. Disclosure of use of natural information in monopoly intellectual property (only a YES or NO check is required)

    +

4. A royalty levied according to the type of utilization and industry

    +

5. When worthwhile, determination of dispersal of natural information utilized according to geographic distribution of species (using iBOL, GBIF, or any system available)

    +

6. When not worthwhile, retention of accumulated royalties to defray the fixed costs of the system

---

The technical solution for fairness and equity in benefit-sharing is a simple deduction, once two basic truths are accepted. The first is the nature of the object of access. For the purposes of biotechnology and securing value through monopoly IP, genetic resources are natural information. The second truth is that the logic of choice belongs to economics, not law. The defense of the foundational error of Article 2 of the CBD, largely by delegates, NGO representatives and consultants trained in law, has driven ABS in the wrong direction at a distance that can be measured in years, dollars or even species lost to extinction.

If both truths were recognized simultaneously, an equitable and efficient ABS regime would emerge almost immediately. The economics are simple and can even be intuited without much exposure to the intricacies of formal economics.

Marrero-Girona and Vogel have reduced it well into one paragraph, albeit dense:

> Inasmuch as genes are information – a sequence of nucleotide bases that can be copied – the analogy with intellectual property is really a homology. Conservationists cannot recoup the opportunity costs of conservation if anyone can trade freely in the same natural information, usually geographically dispersed. Why conserve a vast habitat if you can take out a few samples? Oligopoly rights over natural information are the analog to the monopoly rights over artificial information. Such framing of ABS also extends to enforcement. Similar to artificial information, the illicit flow of natural information cannot be impeded physically. The fence around information must be metaphorical, i.e. a legal instrument. So the economics-of-information narrative ends with analogous institutions: intellectual property has TRIPs [Trade Related Intellectual

Property Rights] and WIPO [World Intellectual Property Organization]; genetic resources should have an International Regime on ABS under the Secretariat to the UN CBD.

(Marrero-Girona and Vogel 2012: pp. 55–65)

Inductively, the shortcomings of bilateral ABS have resulted in an opening for the solution: Article 10 of the Nagoya Protocol envisions a GMBSM and Article 11 on shared resources invites "bounded openness".[7] Deduction would have been a short-cut, saving years, dollars and perhaps a whole ensemble of species lost to land conversion.

The expression "studiously ignored" is both descriptive and emotive. It captures well the resistance of ABS stakeholders to the academic literature as well as a certain exasperation (Vogel *et al.* 2011). As a phenomenon it cuts across a large swath of disciplines. Omar Oduardo-Sierra cites prominent scholars in different fields of work (e.g. G. Stigler, H. Daly, E.O. Wilson) who lament that their contributions to their respective fields were ignored by their peers for needlessly long periods of time, until they were vindicated (Oduardo-Sierra *et al.* 2012).

The same seems to be true for the economic approach to ABS. For almost half a century, a distinguished and peer reviewed literature has existed on the policy implications of the economics of information. Similarly, the explicit recognition of genetic resources as information is just as old (see Chapter 1). A detailed application of the economics of information to genetic resources as natural information has been developed over the last generation and can be traced to the independent works of Swanson, who argues by way of analogy, and Vogel, by the stronger criterion of homology.[8]

Some legislations have indeed recognized the informational nature of genetic resources (e.g. Brazil, Costa Rica) yet fail to develop the subsequent logical regulatory framework of that recognition which would be based on the economics of information. Likewise, some authors and scholars make reference to genetic resources as information but stop short of fleshing out the political consequences for the CBD process, COPs and ABS trajectory in general (Stone 1995; Swanson 1997; Winter 2013; Drahos 2014).

Referencing Vogel's work, others have also advocated cartel-based conceptual frameworks (Drahos 2014: p. 147; Winands and Holm-Muller 2014) while somewhat surprisingly suppressing the essential argument, the justification of rents, because the object or subject matter is information. "Genetic potential", "information component of genetic resources", "intellectual complement [to the material substratum]" and so on are a series of notions which circumvent a dimension in genetic resources first suggested in 1991: that genetic resources are *natural information*.[9]

Oduardo-Sierra *et al.* (2012) tracked the presence of the theoretical framework for the economics of information in ABS literature and CBD related forums. The results were startling: when searching in Google and Google Scholar for hits on "Convention on Biological Diversity" and "access", "benefit-sharing", "openness" and "information", hits are substantial and go into the millions. When "economics of information" is added to the search, results drop to almost nil, other than those related to Vogel, Vogel et al. and a few others. The dearth is surprising. Land use and habitat conservation belong to the discipline of economics and as stressed repeatedly, genetic resources are essentially natural information for the purposes of R&D. How and why did almost everyone fail to connect the two?[10]

Oduardo-Sierra et al. concede the possibility that during the first ten years (1993–2001), lack of due diligence may be an explanation. Researchers and scholars missed both the literature and the connection though one would think that many would have converged on it. The Nobel Memorial Prize in Economics in 2001 was on the economics of information and the "omics" revolution well under way.[11] For the next ten-year period (2002–2011), Oduardo-Sierra et al. suggest the "economics of information" was "studiously ignored", recognizing a high bar in attributing intent. To paraphrase, the failure to mention the implications of the economics of information was, likely, intentional: "the lack of due diligence is not a credible explanation for failure to apply the economics of information in the CBD once one recognizes genes as information" (Oduardo-Sierra et al. 2012: p. 2).

One could add a couple of other reasons for this surprising circumstance. By 2011 and onwards, ABS trends were firmly established and questioning bilateralism, PIC and MAT was, simply, out of the question.[12]

Confirmation of the bilateral approach to ABS in the Nagoya Protocol made it even more difficult to recognize the fundamental flaw of the CBD and embark on the implications of the economics of information. In simple terms, it was and remains politically incorrect to suggest errors in the origins.

In 2007, Vogel began to grapple with the non-reception to the application of the economics of information to ABS. Like the application itself, his thinking was fairly standard economics: interests were vested in the error and the problem was of principal-agent. Significant royalties would result (the vested interests opposed) from new biotechnologies (using natural information) while transaction costs would be substantially lowered (eliminating the rationale for agents). Lower transaction costs means by definition fewer legal, administrative and procedural complexities, which would imply less need for legal experts, consultants and formal bureaucracies to solve these complexities.

A recent example from law firm Wilmer Hale is quite illustrative of what is at stake. Law firms are already craving the opportunities of providing legal counseling regarding the "complexities" (transaction costs) of ABS:

> With the entry into force of the Nagoya Protocol, companies and researchers utilizing genetic resources occurring in foreign countries are subject to *new requirements* governing access and benefit sharing. These *restrictions* raise an *array of legal questions* cutting across distinct areas including compliance, international trade, international litigation, *transactional and licensing arrangements*, and protection of intellectual property rights. Wilmer Hale Attorneys have access to many of the foreign laws adopted by countries to implement the Protocol and the

firm has assisted clients on matters arising under this new framework, both at
the international and domestic levels ....

(Italics added)[13]

The proximate causation of selfish intent may be buttressed by an ultimate causation
in human evolution for group defense. Vogel has modified his original explanation
that resistance springs from just vested interests and the principal-agent problem. He
now believes that the groupthink is an expression of human eusociality and perceives
the need for bold leadership in ABS (Vogel 2013). A natural experiment in "studied
ignorance" and "groupthink" unwittingly occurred from April 8 to May 24, 2013.
The CBD Secretariat organized an Online Discussion of ABS experts to undertake a
broad consultation on Article 10 of the Nagoya Protocol and the need for and modalities of a global multilateral system on ABS. The discussion also addressed Article 11 of
the Protocol. Elsewhere, the Executive Secretary had called for feedback to the CBD.
The mandate was clear: discuss and where possible back interventions with "credible
sources of information, preferably from peer reviewed articles."[14]

In the context of discussing genetic resources shared by more than one country and transboundary genetic resources, or resources for which PIC is not possible, an opportunity arose to test the receptivity of the application. Paraphrasing
Theodosius Dobzhansky's most famous phrase "Nothing in Biology Makes Sense
Except in the Light of Evolution," Vogel tried to make sense of ABS in the light
of the economics of information and explore "the strength and acceptance of the
theory" as suggested by Dobzhansky.[15]

Vogel, Pierre du Plessis and myself brought to the attention of participants the
essential flaws in the current ABS frameworks. We stressed the foundational flaw
of defining genetic resources as material and allowing bilateralism through an
extremely biased understanding of sovereignty as a way to realize benefits as one
would with any tangible good such as timber, fisheries or minerals. Specific and
detailed responses were provided to the interventions of each participant and cita-
tions of peer reviewed papers as requested by the Secretariat supported the position
of "bounded openness".

Tellingly, the 17,500 word Synthesis Report of the Online Discussion produced
by the CBD Secretariat made no reference to the economics of information, nor
to the category mistake by the CBD of defining "genetic resources" as "material".
Although one full page reviewed the policy implications of "bounded openness"
(see Chapter 5), it was followed by another page of detailed objections, all of which
were answered in the forum.[16] The Secretariat would not brook either the category
mistake or the connection. Such path dependency coheres to legal thinking as
expressed in *stare decisis*, Latin for "to stand by decided matters." For policymakers
and lawyers alike, precedent becomes the driving force behind arguments and decisions, which contrasts starkly with economic thinking of "sunk costs", "unrecover-
able past expenditures, [which] should not normally be taken into account when
determining whether to continue a project or abandon it, because they cannot be
recovered either way."[17]

Summarizing, the resistance to any redefinition of genetic resources as natural information and to change in direction away from bilateralism is steadfast. Any move away from MAT and PIC is usually dismissed and, when pressed, distorted as the official Synthesis to the Online Discussion on Article 10 evidences.[18]

One final possible explanation may lie in the proclivity of lawyers and most actors involved in ABS to adhere to *stare decisis* – let the precedent stand – and the subtle dominance of lawyers at all the COPs and in many ABS forums. Indication that the flaw of the CBD is foundational is the very number of COPs – twelve – since 1994. Many conversions of land use also transpired largely because incentives for conservation were and are illusory. Bilateral ABS policy has failed and as the old adage goes, when all else fails, try the truth. The truth begins with a correct classification of the object of access for the purposes of R&D, i.e. natural information. It continues with the recognition that in most cases natural information is dispersed across jurisdictions and culminates with a justification of a GMBSM as an expression of sovereignty.

Neither the CBD nor the Nagoya Protocol are carved in stone. We need creative, innovative and sound policy and legal approaches that provide incentives for habitat conservation, through, among others, the realization of the fairness and equity dimensions in ABS. Reversing a trend, as difficult as it is, becomes a must for the simple reason that, unlike the CBD and the Nagoya Protocol, extinction is irreversible.

NOTES

1 For the seminal article on path dependency in economic theory, see David, P. (1985) "Clio and the Economics of QWERTY." *American Economic Review* 75(2): 332–337.

2 The IP regime took centuries to evolve and consolidate. ABS is new ground. Are we hasty in pushing reform? I would say not inasmuch as a species is lost every 20 minutes according to the estimates from biologists. We cannot wait for another 20 years to ensure ABS regimes respond to human needs and effectively contribute to conservation and sustainability. See Wilson, E.O. (1992) *The Diversity of Life*, 280. New York: W.W. Norton & Company.

3 By "a broad set of actors" I mean: funding agencies, CBD negotiators, indigenous peoples' representatives, ABS consultants, researchers and NGOs. All play an important role in setting the course and bearings for ABS development, in terms of specific policies and regulatory frameworks.

4 Krugman, P. (2014a) "Why Economics Failed." *New York Times*, May 1 (available at http://nyti.ms/1kz4iZ7) and Krugman, P. (2014b) "Point of No Return." *New York Times*, May 15 (available at www.nytimes.com/2014/05/16/opinion/krugman-points-of-no-return.html?_r=1).

5 Krauss, L. (2013) "Deafness at Doomsday." *New York Times*, January 15 (available at www.nytimes.com/2013/01/16/opinion/deafness-at-doomsday.html?_r=0).

6 Lyle Glowka, Susan Bragdon, Brendan Tobin, Charles Barber, John Mugabe and Gudrun Henne acknowledged early in the CBD process the need for a better understanding by the legal community about the intricacies and complexities in the science and technology of genetic resources. Though training and capacity building courses in ABS are plentiful, lacking is a course for non-scientists on the underlying science and more recent developments in genetic resources R&D.

7 The African Group proposed the notion of a multilateral fund during COP 10 as "a new and innovative financial mechanism that would put in place a multilateral approach *in parallel to the dominant bilateral ABS modalities currently being negotiated*" (italics added). The resulting text elaborated the idea and included the notion of a *mechanism*, rather than an isolated funding scheme. See Berne Declaration, Brot, ECOROPA, TEBTEBBA and TWN (2013) *Nagoya Protocol on Access to Genetic Resources and the Fair and Equitable Sharing of Benefits from their Utilization. Background and Analysis.* Penang, Malaysia. p. 76.

8 Two short articles appeared in the newsletter of CIRCIT, an Australian think tank – Vogel. J.H. (1990) "Intellectual Property and Information Markets: Preliminaries to a New Conservation Policy." *CIRCIT Newsletter* (May): 6 and Vogel, J.H. (1991) "The Intellectual Property of Natural and Artificial Information." *CIRCIT Newsletter* (June): 7 – which set the basis for the conceptual development of applying the economics of information to genetic resources as natural information. Tim Swanson (1992) touched on the same themes in a discussion paper entitled "The Economics of the Biodiversity Convention" (CSERGE Discussion Paper GEC 92-08, available at www.cserge.ac.uk/sites/default/files/gec_1992_08.pdf).

9 Circumlocution of natural information is evident in the following passage from Chege Kamau and Gerd Winter:

> An analysis of the sovereignty and property (or 'ownership' which shall be used as the generic term in this article) over the *genetic potential* of a biological resource must take into consideration two possible objects of ownership over the *genetic potential*: the genome being the material substratum and the *information about the genome* being the *'intellectual' complement.*
>
> (Italics added)

See Kamau, C.E. and Winter, G. (2013a) "An Introduction to the International ABS Regime and a Comment on its Transposition by the EU." *Law, Environment and Development Journal* 9(2): 108–126. Available at http://ssrn.com/abstract=2387876

10 For a detailed overview of this survey and its results, see Oduardo-Sierra, O., Hocking, B.A. and Vogel, J.H. (2012) "Monitoring and Tracking the Economics of Information in the Convention on Biological Diversity: Studied Ignorance (2002–2011)." *Journal of Politics and Law* 5(2): 29–39.

11 Oduardo-Sierra *et al.* (2012) ibid.

12 Other authors such as Halewood, Winter, Chege Kamau, Bevis Fedder, De Jonge and Young also question bilateralism from the perspective of "common pools". See, for example, Kamau, E.C. and Winter, G. (eds.) (2013b) *Common Pools of Genetic Resources. Equity and Innovation in International Biodiversity Law.* London and New York: Earthscan from Routledge, or Halewood, M., Lopez Noriega, I. and Louafi, S. (2013) *Crop Genetic Resources as a Global Commons. Challenges in International Law and Governance.* Issues in Agricultural Biodiversity. Abingdon and New York: Bioversity International, CGIAR, Earthscan from Routledge.

13 See article by attorney of Washington DC based firm Wilmer Hale: Manheim, B. (2014) "Nagoya Protocol Spurs New and More Stringent Requirements for Prior Informed Consent and Benefit Sharing for Research and Commercial Activities Involving Genetic Resources from Plants, Animals and Microorganisms." October 17, 2014. Available at www.mondaq.com/unitedstates/x/347698/Life+Sciences+Biotechnology/Restrictions+Governing+International+Trade+in+Genetic+Resources+Enter+Into+Force

14 CBD Secretariat. Notification No. 2013–018 (2013), available at wwwcbd.int/doc/notifications/2013/ntf-2013-018-emerging-issues-en-pdf

15 Dobzhansky, Theodosius (1973) "Nothing in Biology Makes Sense Except in the Light of Evolution." *The American Biology Teacher* 35: 125–129.

16 The complete Online Discussion may be compared with the official synthesis report to assess the faithfulness of the representation. Available at https://bch.cbd.int/abs/art10_groups/searchforum/

17 See http://economics.about.com/od/economicsglossary/g/sunkcosts.htm

18 During the online debate, 47 interventions cited economics and 22 interventions the economics of information although none appeared in the Synthesis, nor did the supporting academic works despite the call in Notification 2013-018 of the Executive Secretariat of the CBD to cite credible sources of information and peer reviewed articles in particular. The Synthesis also referenced *en passant* the argument for genetic resources as natural information despite many interventions to the effect. See UNEP/CBD/ICNP/3/INF/4 Synthesis of the Online Discussion on Article 10 of the Nagoya Protocol on Access and Benefit-sharing, April 2014, available at www.cbd.int/doc/?meeting=ABSEM-A10-01

Chapter 5

"Bounded openness" as fair, equitable and efficient

The path to a global multilateral benefit-sharing mechanism

The previous chapters suggest a dozen questions about ABS: internalizing the answers in the COPs will put the CBD on a path to achieving its three objectives.

1. What are the implications for the CBD of redefining "genetic resources" as "natural information"?
2. What does "country of origin" mean when the geographic distribution of natural information occurs across national borders and taxa?
3. How can one interpret "sovereignty" in light of the answers to (1) and (2)?
4. Are "bounded openness" and "common pools" synonymous?
5. How can "bounded openness" be operationalized?
6. Can advantages be tabulated?

WHAT ARE THE IMPLICATIONS OF REDEFINING GENETIC RESOURCES?

The primary implication of a correct classification of genetic resources is that the economics of tangibles will not be efficient, fair or equitable, when applied to intangibles. The secondary implication is that a sole "country of origin" is only an accurate description for the limiting case where the natural information is found only and exclusively in one jurisdiction. Competition among different countries of origin which possess the same natural information will eliminate rents as the price of access falls to the marginal cost of collection, which is essentially nothing. For an efficient, fair and equitable ABS policy, sovereignty must be reinterpreted as the right of countries to join a regime that fixes a price and distributes rents to the different "countries of origin".

The criterion of efficiency for ABS policy must also reflect the very low probabilities associated with commercially successful IP that derives from utilization of natural information and the cognitive bias to confuse the mathematical expectation of the probability multiplied by the payoff, as if it were a certainty.[1] Distributing rents from commercially successful IP that derives from natural information (particularly but not only patents) makes both accounting and common sense. Hence, openness in access must be "bounded" by disclosure requirements on IP claims to enable to charge for and distribute the royalty benefits. Patents have been the primary focus for the obvious reason that they involve the greatest values added to natural information but, in principle, the proposal of bounded openness could be applied to a wide range of IP (see Case Study 2).[2]

The political scientist Christopher May introduced the concept of "bounded openness" in the general context of IP regimes. May has been highly critical of the unbalanced nature of IP and perceives that private interests are highly favored over the public good in policymaking.[3] He exemplifies openness in the case of Wikipedia and certain software applications and scientific research – all of which simultaneously impose certain accessibility limits and boundaries. Bounded openness is described as "a balance to the closed logic of ownership" (May 2010: p. 7).

Vogel adopted May's generic term to define the legal orientation of a new international ABS framework or GMBSM under which openness means that "genetic resources [natural information] would flow freely with some notable exceptions (e.g. endangered, invasive and pathogenic species)." The bounded[ness] dimension means that a royalty is imposed on commercially successful IP and distributed among countries of origin (of species or sub-species "providing" natural information) proportional to habitat where this information is maintained (Vogel 2012: p. 184). Patents and plant breeders' rights, would be the most logical IP tools at first glance.

Countries would be exercising their sovereignty by placing their natural information within the contours of a mutilateral system which will reward them with money to conserve species and habitats or spend on projects that render the highest social return. Such a regime would be an exercise in sovereignty not unlike any other international convention. One must also remember that often money is not needed to conserve habitats but to offset the political pressures to change land use. The distinction is nuanced and mostly lost in the COP.

WHAT DOES "COUNTRY OF ORIGIN" MEAN WHEN THE GEOGRAPHIC DISTRIBUTION OF NATURAL INFORMATION OCCURS ACROSS NATIONAL BORDERS AND TAXA?

With respect to "genetic resources that occur in trans-boundary situations" (Article 10 of the Nagoya Protocol), no consensus exists as to the meaning of a transboundary situation. From the reductionist view of genetic resources as natural information, most of the objects are transboundary, widely dispersed across jurisdictions and, counterintuitively, across taxa (e.g. the same active compounds or secondary metabolites found in species separated by genera or families). Others may be so localized as to be found in one population at one moment in time (see Case Study 1). So, the question of diffusion is empirical and must be addressed by the GMBSM. Such an interpretation would also include the more extreme cases of genetic resources in the deep sea bed, high seas and Antarctica.

Article 10 stipulates that "[T]he benefits shared by users of genetic resources … through this mechanism shall be used to support the conservation of biological diversity and the sustainable use of its components globally." Though the phrase "benefits shared by users of genetic resources" is confusing it seems to imply that benefits in general will be shared with Providers and directed at conservation and sustainable use of biodiversity and its components at a global level. The "global multilateral benefit-sharing mechanism" may involve an international binding legal instrument (i.e. another protocol) and a funding mechanism of which a part would be governed by the principles of bounded openness and specifically targeted toward *in situ* conservation and habitat preservation.

Unfortunately, basic economics is again being ignored. The highest social return may be not in a conservation investment but, say, in watershed projects, literacy campaigns or vaccination. If the agricultural frontier is not expanded and the means of transport cumbersome, biodiversity can be conserved with very little expenditure. Remote areas in the Amazon basin are a good example. The most cost-effective conservation measures are incentives to stop and prevent action, such as changes in land use or river dredging rather than restoring degraded habitats. Similarly, the question of fungibility looms large; if the benefits generated from ABS are directed toward priorities which would have been funded anyway, the incremental effect will be nil.

Article 10 is often interpreted strangely in order to let the precedent stand on bilateralism. Thus, the transboundary situation is cast as an exceptional possibility. This is highly ironic given resources in transboundary situations are a common occurrence and that bioprospecting in the high seas and deep sea bed is also becoming more common.[4]

Hope lies in the legal text that a "special" multilateral regime will be required. Despite the caveats on allocating financial benefits to the highest marginal social benefit and concerns about fungibility, the Nagoya Protocol expressly links ABS directly to global *in situ* conservation and sustainable use.[5] This becomes especially relevant for a GMBSM based on bounded openness which could align incentives for conservation while avoiding both problems.[6]

Article 11 should be read in conjunction with Article 10 and interpreted to support a "global multilateral benefit-sharing mechanism." Article 11(1) establishes that in cases "where the same genetic resources are found in situ within the territory of more than one Party, those Parties shall endeavour to cooperate, as appropriate, … with a view to implementing this Protocol."

Epistemologically, "the same" genetic resource is tacit recognition that "genetic resources" are information, as a piece of matter cannot be found in two or more places at the same time. This did not go unnoticed during the Online Discussion on Article 10 of the Nagoya Protocol. Most participants sided with the view that contracts are possible and efficient, no matter whether materials with similar properties are found in and obtained from different sources.[7] Precedent also exists in the notion of "range states" for migratory species (De Klemm 1994).

Developing a GMBSM requires political will "to endeavour or cooperate." Despite the realpolitik of ABS and the institutional resistance to change, many quarters do evidence willingness to revisit long-standing assumptions in ABS. A mandate for a full-blown reformulation may not yet have arrived, but the notion is no longer unimaginable.

The country of origin or Provider is relevant in terms of the initial determination of the species used as sources of natural information during phase one of any bioprospecting project involving *in situ* collecting or *ex situ* sampling.[8] The country of origin will need to grant permits and permissions for collection of samples in a sustainable fashion, or negotiate the non-monetary benefits[9] and impose a series of obligations on applicants (Users) – regarding compliance with national administrative procedures. By understanding that a country of origin is not sole proprietor (others possess the same natural information), administrative access procedures should be significantly streamlined and *facilitate* collecting, bioprospecting and R&D. This in no way undermines sovereignty.

If the concept of "genetic resources" gives way to the notion of natural information, the object of access could read: "any information, derived from nature, but not limited to, hereditary units, metabolites, proteins, enzymes, prions, phenotypic expressions and non-human cultures." Hence, the focus of policies and regulatory frameworks automatically translate to the *utilization* of the natural information.

HOW CAN ONE INTERPRET "SOVEREIGNTY" IN LIGHT OF THE ANSWERS TO (1) AND (2)?

Sovereignty requires a different interpretation. Countries or parties have the right to define how to participate fairly and equitably in the monetary benefits derived from access to and use of natural information, including through the development of a GMBSM. The GMBSM would be a cartel-like structure which distributes royalties collected from the commercially successful uses of natural information that carry intellectual property protection. How Providers ultimately use the money should vary according to the highest social return; nevertheless the percentage shared diminishes should a country allow the habitat of its species to shrink. Incentives are thus aligned.

The approach would not preclude countries from defining bilateral contractual relations with Users to define non-monetary benefits derived from access to and use of the vehicles of natural information – the specimens from different species of plants, animals, microorganisms and other biological entities. The long-standing practice of permits and authorizations would remain but for reasons other than ABS or PIC, say, to prevent genetic erosion or reduce the introduction of exotic species. For example, Users may need to pay certain fees to collect in protected areas – calculated on number of specimens collected, weight of biological material extracted, or other criteria – where the revenues offset the costs of protection.

How does one know that natural information has been successfully utilized in R&D? Simple disclosure would be required and would vary according to the intellectual property sought. For patents, a simple check on the application, for copyrights and trademarks, a variant on the copyright and trademark symbols, and so on down the list of intellectual property categories. In theory, bounded openness could apply to any form of IP. However, one would start with the most valuable categories added first, which lie in biotechnological patents and maybe certain plant breeders' rights. Applications of bounded openness for other categories of IP are left for future elaboration.

Due to the category mistake of considering genetic resources as material, the response for the need of disclosure[10] has also been somewhat distorted. The concept of "certificates of origin" has been on the ABS agenda for almost two decades. Brendan Tobin was the first to use the concept in the context of a standard document which discloses legal access to genetic resources and TK and which "follows" resources along the R&D chain (Tobin 1997).[11]

Vogel, however, finds a certificate of origin an unnecessary transaction cost, especially needless given that most R&D does not result in commercially successful intellectual property. He proposes that a one-line disclosure system is used to identify the species in the patent application (Vogel 1997). After further reflection, he believes that greater efficiency can be had through just a disclosure of whether or not natural information was used, without identification of the species in the patent application to maintain confidentiality of research streams at that moment. Should commercial uses ensue, the system would require a follow-up disclosure of the species.

As mentioned previously, identification of the species may not be an immediate and straightforward affair: a transition period may be needed until the moment that identification of the species or appropriate taxonomic level can be justified. Identification would be deferred (in some cases) to the time of commercial success. Bounded openness allows for that flexibility.

Today the notion of "certificate of origin" has become mainstream in ABS and recognized in the Nagoya Protocol as a Certificate of Compliance – but Vogel's suggestion, albeit far simpler and less cumbersome, is strangely viewed as radical, perhaps because it ties into a very different alternative ABS approach.

Building on the general idea of certificates (e.g. certificate of legal provenance, certificate of compliance, certificates of origin, etc.), lengthy reports and studies have been prepared over the years, taken on board ultimately in Article 13(3) of the Nagoya Protocol, which determines that:

> [A]n internationally recognized certificate of compliance shall serve as evidence that the genetic resource which it covers has been accessed in accordance with prior informed consent and that mutually agreed terms have been established, as required by the domestic access and benefit-sharing legislation or regulatory requirements of the Party providing prior informed consent.

The notion of "certificate of compliance" corresponds to bilateral negotiations and PIC, and seeks to prove that a User complied with national PIC and MAT conditions. To reduce transaction costs to a minimum, the certificate would simply indicate: date of issuance by a national authority, country providing the specimen and the species or subspecies (if known at the time) or an obligation to include this determination in the certificate as soon as taxonomy defines classification. All of this should be available online, maybe even through the ABS Clearing House Mechanism of the CBD. This dimension coincides with the existing operational framework under the Nagoya Protocol.

At the moment of lodging a patent application, the trigger for disclosure is activated. Like the concept of "certificate", much has been written on disclosure in the patent regime.[12] Legal and policy analysis abounds, both supportive and critical. The idea of disclosing the legal access to a genetic resource or TK can be traced to the Andean ABS process of the early 1990s. Specific legislation with mandatory disclosure was enacted in 1996 linking ABS to IP and the patent regime.[13]

In contrast, Vogel's suggestion eliminates the transaction costs of all certificates. An amendment would be needed in international patent law to require disclosure of whether or not natural information was utilized in the R&D of the invention for which the patent is sought.[14] But this would basically mean that most patented biotechnologies would not disclose the origin of the natural information used during their development as most are not commercially successful.[15] In a regime of bounded openness, detective work is *ex post* and is financed by set royalties accumulated in escrow.

Attempts have been made to modify international patent law, specifically TRIPS, with no success. Efforts to streamline disclosure into IP have a long-standing history, going back to when the Committee on Trade and the Environment of the WTO and during the TRIPS review process under the Doha Round began discussing the matter. At the national level a few countries, including developed countries such as Norway and Switzerland, have already amended their IP legislation, requesting disclosure – albeit for PIC and MAT.

Judgment would be exercised to estimate the cost for a reasonably accurate assessment of geographic distribution. If the estimate is high, royalties will have to accumulate in the escrow until a threshold is met. Suppose, for example, it takes Z to determine the estimated diffusion of a species W and over the 20-year period of the patent and all that accumulates for the patented product that derives from W is (Z + \$1). Should Z be expended on determining geographic distribution to divide \$1? Obviously not! Probably, the threshold of accumulated monies will have to be some multiple of Z – for example, if it costs US\$200,000 to determine distribution, then the threshold could be $2Z$ with US\$200,000 to be distributed and US\$200,000 to cover the costs of determination. If over the 20 years the sum only comes to US\$300,000 then it makes more sense to defray the costs of the inventory the full US\$300,000 rather than spend US\$200,000 only to be able to disperse US\$100,000. Probably, the most acceptable for all parties would be $2Z$, based on cognitive biases for 50–50 splits. In other words, the sum to be distributed must be more than the costs entailed in making the determination of geographic distribution (Figure 5.1).

Decision Tree of Disclosures and Benefit Disbursement

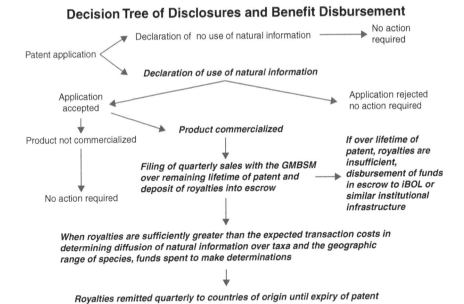

Figure 5.1 Disclosure simplified

A difference between bounded openness and bilateralism is that under the latter, disclosure is a condition for either processing or granting of a patent, and its absence can invalidate the patent.[16] Under bounded openness (Figure 5.2), minimal disclosure is a requirement for processing the patent application, thereby encouraging R&D. So, if an applicant uses natural information but does not disclose for whatever reason in the application, the patent would not be revoked once granted. Rather, penalties would be institutionalized to encourage truthful revelation whereby the probability of the non-compliant User being detected multiplied by the penalty is significantly greater than the expected royalties due. Incentives could thereby be better aligned for both R&D and disclosure.

The new ABS international regime will require determination of the geographic distribution of species around the world. Databases have been developed with computational models which provide critical data regarding location and distribution of species, especially in certain cases such as marine species, higher plants and mammals. The International Barcode of Life (iBOL) is a powerful tool to determine distribution, assess evolutionary patterns and undertake ecological studies (Box 5.1). The Global Biodiversity Information Facility (GBIF) likewise generates very precise information and data which can assist in determining geographic distribution of species.[17] Both could act as enablers of a new international ABS regime, although neither was ever so intended.[18]

An important point must be stressed: bounded openness does not require precision regarding the countries of origin. As technology advances and ecological knowledge accumulates, greater and greater precision will arise. Vogel argues that even the old Holdridge zones[19] could have worked, crudely but well enough, at the outset of the CBD in 1993.

Box 5.1 What is iBOL and why is it important?

Identification and discovery

iBOL uses sequence diversity in short, standardized gene regions – DNA barcodes – as a tool for *identifying known species* and discovering new ones. By reinforcing traditional taxonomy, DNA barcoding is revolutionizing our capacity to know and monitor biodiversity.

The iBOL mission

iBOL's main mission is extending the *geographic and taxonomic coverage* of the barcode reference library – Barcode of Life Data Systems (BOLD) – storing the resulting barcode records, providing community access to the knowledge they represent and creating new devices to ensure global access to this information. That includes a hand-held device that will provide real-time access to identifications by anyone in any setting.

Real world problems

iBOL researchers will also work on applying DNA barcoding to real world problems, such as forensics, *conservation*, marketplace regulation, control of diseases and *ecosystem monitoring*. Within five years, iBOL participants will gather DNA barcode records from five million specimens, representing at least 500,000 species. This will produce an excellent *identification system for economically, socially or environmentally important species* and a solid foundation for subsequent work on barcode reference library for all life.

Wide-ranging impacts

Once implemented, this DNA-based identification system will exert broad impacts on all areas in which society interacts with biodiversity – pest and disease control, food production and safety, *resource management, conservation, research*, education and recreation.

Economic benefits

The economic benefits of improved bio-surveillance will be large. Increasing globalization of trade and climate change means that all jurisdictions face unprecedented exposure to invasive species that threaten their agriculture, forestry and fisheries. DNA barcoding will enable the rapid identification of invasive species, allowing quarantine and eradication efforts to begin far earlier, with massive reductions in cost and increased chances of success. It will further aid the selection of optimal control strategies for pest/disease agents impacting all natural resource sectors. Barcoding will play a critical role in regulating trade in endangered or protected species and products.

Early warnings

As massively parallel sequencing technologies become more available, the barcode library will enable sophisticated *environmental monitoring* that uses living organisms as integrators of environmental change and as early warnings of damage. Large-scale, automated monitoring of species presence and abundance in the world's oceans, inland waters, agro-ecosystems and plantations will soon be routine.

Research alliance

iBOL's work will be carried out by a research alliance spanning 26 nations with varying levels of investment and responsibilities. The overall task of iBOL researchers is collecting and curating specimens, gathering barcode records from them and building the informatics platform needed to store these records and to enable their use in *species identification*.

A central role

Because Canada has led the early development of this DNA-based identification system, it will play a central role in iBOL. Research involvements will span the nation, but the Biodiversity Institute of Ontario (BIO) and its staff of 100 genomicists, informaticians and taxonomists will be iBOL's scientific hub, operating its high-volume sequencing facility, maintaining its central informatics platform and hosting its largest scientific team.

Source: iBOL website (http://ibol.org/). Italics have been added to potential areas of synergy between iBOL and bounded openness.

The economics is simple and corresponds to Chapter 8 of the classic textbook *Economics* (Samuelson and Nordhaus, 2005). Competition is dynamic and each supplier in a competitive market has an incentive to provide the good, as long as the price is above its marginal cost. In the case of genetic material, the Providers are countries and competition will be keen whenever the source of the genetic material, namely, a species, overlaps jurisdictions. Competition reaches its zenith when the same genetic material can be obtained in the United States, a megadiverse Provider and non-Party to the CBD. In the US, the attribute of interest in the genetic material, its information, is legally the property of no one (*res nullius*). Try selling something when someone else is giving it away!

Even for natural information that cannot be had for free in the US, a price war among other Providers makes it almost free. Rents are quickly eliminated for any

Provider that wishfully thinks the User should mitigate the opportunity costs of habitat conservation, i.e. paying a premium for not felling the timber where the leaves were collected, for not damming the river where the microorganisms were drawn and so on. So, without a *sui generis* system that fixes the royalty rate and a scheme for the distribution, no rent whatsoever can be realized through an ABS contract or an MTA Industry knows the economics first-hand. Once a patent expires on a biotechnology, generic manufacturers enter the market and the price drops sometimes by as much as 90 percent. Economists, however, do not criticize the original monopoly price. They make an exception to their general rule pro competition in the case of IP, which is artificial information. They recognize that the fixed costs of R&D must be recouped over the life of the patented product. By extension, they should also recognize that rents must be institutionalized through a *sui generis* system on natural information. The opportunity costs of conservation must be addressed to discourage land use conversion.

ARE "BOUNDED OPENNESS" AND "COMMON POOLS" SYNONYMOUS?

Stephen Brush (1996) warned that the concept of "commons" should not be conflated with "open access", recalling the unfortunate title of Garrett Hardin's classic article ("The Tragedy of the Commons"). A corollary exists to his prescient warning for the terms "common pools" and "bounded openness", which also invite confusion.

The commons has been much celebrated in the economic literature as evidenced by the award of the 2009 Nobel Memorial Prize to Elinor Ostrom, who observed that common pool resources are indeed managed as if common property resources, based on auto-regulation. The body of her work seems to disprove the premise in the title of Hardin's article. However, language is a matter of choice. What Hardin meant by commons was "open access" and not jointly held property. Regulating assets through rules was one of the alternatives suggested in his classic but mistitled article.

In the context of ABS, the "common pools" of Kamau and Winter would require some level of negotiations and formal and informal agreements to define responsibilities and obligations for the commoners. It does not spring from open access, nor does it converge on the recommendation of "bounded openness".

Most scholars agree that it is impossible to enclose artificial information, and natural information is no different. De facto, natural information will remain open and biopiracy easily executed. However, exclusivity for value added to natural information is possible through the IP system. Under bounded openness, inventors would disclose the natural information used in pursuit of a monopoly intellectual property. One would not need to bother with determining *ex ante* who are commoners if the value added does not result in a commercial product.

In the few cases where there is a commercially successful biotechnology, only then does one identify all the countries of origin which could be commoners.

Note that in the Kamau and Winter scenario, we essentially would have bargaining between the administrators of the commons (Providers) and the User. Although transaction costs would be lowered through economies of scale among Providers, they would still constitute a significant and unnecessary burden, especially if R&D over natural information is not commercially successful. In bounded openness the bargaining has already been institutionalized in a schedule of royalty rates, dependent on a combination of characteristics in the use of the natural information (e.g. direct, indirect, type of industry, and so on). Once operative, the transaction costs of ABS are essentially eliminated.

HOW CAN "BOUNDED OPENNESS" BE OPERATIONALIZED?

An oligopoly over natural information or biodiversity cartel can be intuited in the GMBSM under bounded openness. With the operationalization of Articles 10 and 11 of the Nagoya Protocol there is no need to form a cartel independent from the CBD. A harbinger of the idea was the Group of Like-Minded Megadiverse Countries (GLMMC) which perceived the need for some sort of multilateral benefit-sharing mechanism. In the end, the Group proved incapable of effecting a mechanism as they abided by an interpretation of sovereignty that permitted only bilateral negotiations.

A second element of the operational framework involves a funding mechanism – an existing or to be created scheme – which is responsible for distributing monies from the royalties received. Which country receives what share depends on geographic distribution, for which iBOL and GBIF seem particularly well suited (see Box 5.1). The percentage division proportional to habitat seems the fairest and most equitable as well as the most efficient.

The third element to consider is the coordination of the GMBSM with national regulations governing collection of biological materials and samples to avoid endangerment of species and phytosanitary and pest-species concerns. Most countries have some form of system already in place. In many cases, these same regulatory systems also define forms of non-monetary benefit-sharing in terms of technology transfer, participation of nationals in research, type sample deposits, and so on.

The fourth element is monitoring and tracking of commercially successful patents which have disclosed the utilization of natural information, by ticking Yes in the Yes/No option in the patent application. The patent holder makes a quarterly payment of the royalty owed to the benefit-sharing fund. Every Party of the Protocol would oblige national institutions to adapt and comply. An audit would verify that the correct royalties are transferred into the fund until the patent expires. Royalty rates would vary according to a combination of characteristics in the value-added process. Vogel suggests that:

> A table of flat rates must be negotiated according to the constellation of relevant characteristics. Simple math can elucidate how many cells will have to be filled in the proposed table. There are at least six sectors and ten IP rights suited

to products/services derived from natural information; the presence or absence of substitutability of inputs in R&D or the presence or absence of substitutability of inputs in production, and directness in derivative use or indirectness in research streams.

Theoretically at least 480 distinct cases exist for which the GMBSM can negotiate flat royalty rates. Of that large set, the COP should begin with those combinations that hold the highest mathematical expectation, i.e. the probability of the event multiplied by the value of the event.

(Vogel personal communication 2015)

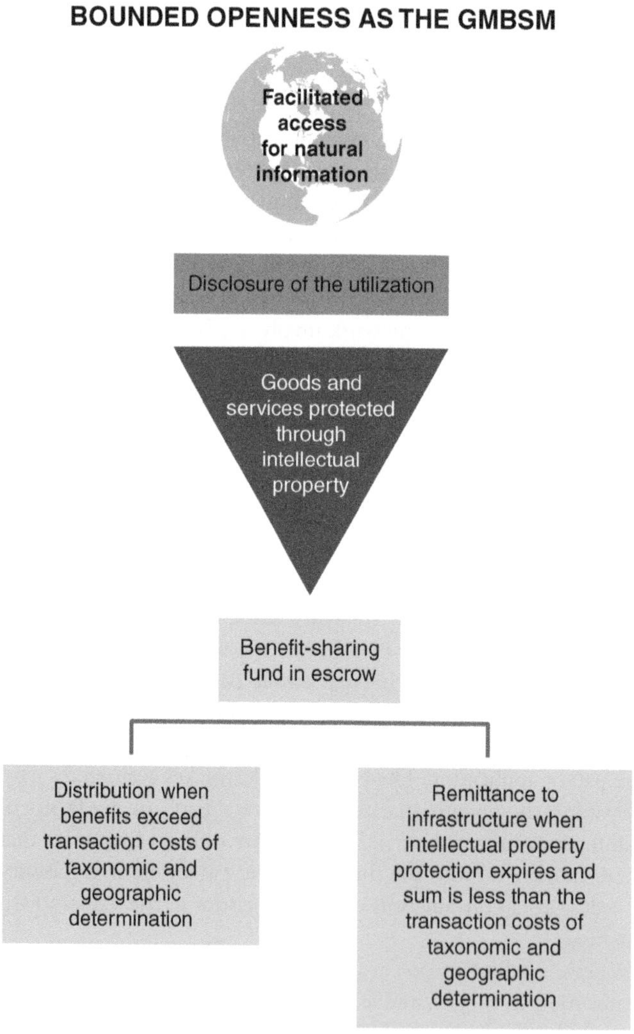

BOUNDED OPENNESS AS THE GMBSM

Figure 5.2 A visual approach to the new global multilateral ABS system

These different elements should be negotiated through an international political process which designs the statutory provisions of the new ABS regime – ideally through COP with inputs from SBSTTA.

CAN ADVANTAGES BE TABULATED?

The overarching advantage for Providers and Users of natural information is that R&D on genetic resources can proceed unencumbered. No new constraints are imposed on collecting biological samples and specimens. There is a quasi-free flow of resources for the purpose of R&D. Under bounded openness, control is established *ex post* if and when monetary benefits are generated. As part of CBD and ABS discussions over the past two decades, emphasis has been placed on the regulatory dimension in ABS *ex ante*, overlooking the principle of facilitating access which is also enshrined in the CBD. Moreover, with bounded openness, the paradox is obviated that few benefits will be generated when access is encumbered.

Cumbersome regulatory frameworks and administrative procedures are now not only having a perverse effect on R&D, they are also inhibiting basic research (Cock *et al.* 2010; Hoagland 1998; Mansur and Cavalcanti 1999; Grajal 1999). Bounded openness will allow for a continued flow of resources for taxonomic and advanced research and development. Even the peculiar case of human pathogens is also resolved (Vogel 2013).

Under bounded openness, the issue of *ex situ* collections and prior or post CBD collections is solved. Just because the material vehicle of the natural information was transferred before the CBD does not mean that all of the natural information therein was accessed. If natural information were obtained from an *ex situ* collection, the claim of Providers would exist as long as the species were not extinct *in situ*. Likewise, if no IP is sought, then there is no obligation as the public domain status of value added is a universally shared benefit.

There is no temporal restriction regarding the movement of genetic resources *ex situ*. Patent filing and disclosure indicate potential for the generation of benefits under the system. Inasmuch as *ex situ* genetic resources are also part of the geographic distribution of the natural information, they are not discriminated and also will have a claim when a genetic resource is extinct *in situ*, again aligning incentives for conservation.

Bounded openness integrates well with the ITPGRFA, and covers those materials not under the Multilateral System and the list of crops (i.e. used for other purposes than food and agriculture).

Benefits are shared only if the intellectually protected innovation (biotechnology) is successful in commercial or industrial terms. Bounded openness is ultimately pragmatic as it wastes no financial resources on commercial dead ends. With the exception of the table of royalties, nothing else needs to be painfully negotiated. The fixed and institutionalized table of royalties is applied to net quarterly sales and the result is transferred to the benefit-sharing fund for proportional distribution among countries of origin of species. This becomes the most equitable and fairest

Table 5.1 All contentious issues under ABS solved and facilitated through bounded openness (v. ABS bilateral approach)

Issues	Under the ABS bilateral approach	Problems/Challenges	Under bounded openness
Country of origin (or Provider)	Grants PIC and negotiates MAT (ABS agreements)	Complex ABS administrative procedures; PIC and MAT at different levels (i.e. indigenous communities)	Irrelevant – except to identify species from which natural information is extracted; Ex post commercialization
Ownership of (genetic resources, derivatives) [natural information]	State has sovereignty, ownership, property, etc. depending on legal status of natural resources in each country	Defining what exactly it is over which State/others have ownership rights	Irrelevant: implicit in the GMBSM
Ex situ materials (genetic resources, derivatives) [natural information] collected prior to the CBD	Depends on the status of collections	Defining legal status of millions of entries in ex situ collections	Irrelevant: the trigger for benefit-sharing is commercial use of the natural information (regardless of when materials were collected)
Ex situ materials (genetic resources, derivatives) [natural information] collected prior to the Nagoya Protocol	Depends on the status of collections	Defining legal status of millions of entries in ex situ collections	Irrelevant: the trigger for benefit-sharing is commercial use of the natural information (regardless of when materials were collected)
Materials (genetic resources, derivatives) [natural information] collected in transboundary situations (outside national jurisdictions)	To be defined – depending on the definition of "transboundary" possibly according to the rules and principles applicable to Antarctica, deep sea bed, and other special jurisdictions. Establishing arrangements for distribution of benefits for conservation of these areas or resources with relevant management entities	Political negotiations may take years in future COPs	Irrelevant: the trigger for monetary benefit-sharing is commercial use of the natural information (regardless of where materials were collected)

Non-commercial (including taxonomic research)	Special contractual arrangements	Determining the boundaries between non-commercial and commercial research in a highly sophisticated R&D environment	Irrelevant: access to materials (genetic resources, derivatives) [natural information] is facilitated for all purposes. Inasmuch as there is no commercial intent through IP there is no disclosure
Changes in the use of genetic resources and derivatives [natural information] (during R&D)	Special contractual arrangements required – conditions + verification mechanisms	Determining the boundaries between non-commercial and commercial research in a highly sophisticated R&D environment and identifying when changes in use occur (often outside national jurisdictions)	Irrelevant: access to materials (genetic resources, derivatives) [natural information] is facilitated for all purposes. Inasmuch as there is no commercial intent through IP there is no disclosure
Multiple sources of genetic resources and derivatives [natural information]	Complex contractual arrangements	Monitoring and tracking complex contracts and multiple R&D activities plus multiple genetic resources from multiple sources	Determination of habitat distribution occurs when economically worthwhile
Materials under the list of the ITPGRFA (for uses other than as PGRFA)	Predefined SMTA determines benefit-sharing (non-monetary and monetary – through set percentage)	Monitoring and tracking complex contracts and multiple R&D activities plus multiple genetic resources	Irrelevant: access to materials (genetic resources, derivatives) [natural information] is facilitated for all purposes
Calculation of monetary benefits	Case by case negotiations	Intrinsic inequities in bilateral approaches to shared resources. Asymmetries in information between Users and Providers. Potential values of genetic resources impossible to calculate ex ante	A table of fixed royalty fees according to characteristics in utilization (e.g. type of industry), when biotechnology has a commercial success. Amount is calculated through an accounting process with companies
Calculation of non-monetary benefits	Case by case negotiations (under current administrative frameworks)	There is a culture and practice in negotiating non-monetary benefits	Case by case negotiations (under current administrative frameworks)
Trigger for benefit-sharing	Case by case negotiations for access of the physical material	Monitoring R&D process (often outside countries). Reliance on good faith	Commercial success of a biotechnology

(continued)

Table 5.1 (cont.)

Issues	Under the ABS bilateral approach	Problems/Challenges	Under bounded openness
Checkpoints	To be defined	IP offices and other to be defined (e.g. commercialization points)	IP offices
Monitoring	To be defined: ABS clearing house mechanism (CHM) will play a role Certificate of compliance	IP offices and other to be defined (e.g. commercialization points) – through CHM and national authorities	Once a patented and commercial product is generated Condition: Need for a mandatory disclosure requirement regarding use of natural information (yes/no) + country where material was accessed + species (if known at the time of application)
Compliance	Certificate of compliance	National legislation needs to regulate on behalf of the interests of Providers	Once a patented and commercial product is generated Condition: need for a universal disclosure requirement
Institutional settings	ABS competent authority	ABS competent authority	ABS competent authority iBOL, GBIF, others, to support taxonomic identification of species (if not known at time of collection or during R&D process) and spatial distribution of species or taxon
Areas beyond national jurisdiction (Antarctica, deep sea bed, etc.)	No solution at present – Nagoya Protocol suggests cooperation or a GMBSM	To be defined by Nagoya Protocol under Articles 10 and 11	Commercial success of a biotechnology (patented) provides monies to the benefit-sharing fund for supporting in situ conservation in these areas or beyond Monetary benefits always shared – How? Royalties can be used to support bounded openness institutional structure (e.g. role of iBOL, GBIF, small secretariat)

Source: Author.

approach to share benefits resulting from utilization of natural information obtained from species i.e. through saps, resins, genes, molecules, samples or any biological derived product.

Bounded openness is also an elegant solution. It is required not only for the third objective of the CBD (ABS) but also for the first and second objectives, on conservation and sustainable use (Table 5.1).

NOTES

1 Even one blockbuster drug like Taxol or game-changing biotechnology like PCR can justify the public investment in an ABS policy framework.
2 Copyright could easily be amended to require some sort of statement of natural information utilization, perhaps a new symbol (niu); trademarks could have an underscore (R), and so on for the other forms of intellectual property that derive from natural information. Although the focus is usually on patents and plant breeders' rights, other forms of IP can also assimilate natural information.
3 Almost a decade ago, the UK Commission on Intellectual Property Rights (2001–2002) warned about the need for a balanced IP system (patents in particular) which took due consideration of developing countries and the needs of the world's poor. Trends at the time, including concentration of patents, extensive and excessively broad patent claims, and their detrimental effect on innovation and erosion of the public good, have since intensified. In 2007, WIPO adopted the Development Agenda of WIPO with a view to placing development (uneven in countries) at the heart of the organization's work (see www.wipo.int/ip-development/en/agenda/overview.html). For a developmental perspective on IP which highlights the often pervasive effects of IP on human rights, food security, access to medicine, R&D, TK, biotechnological innovation, etc., see Wong, T. and Dutfield, G. (eds.) (2011) *Intellectual Property and Human Development. Concerns, Trends and Future Scenarios*. Cambridge: Public Interest Intellectual Property Advisors.
4 Since a seminal article on the policy and legal implications of marine research and deep sea bed bioprospecting was written by Lyle Glowka in 1995, numerous papers and publications have been produced, all coinciding with the increased interest by industry to explore the potential of natural information from the deep sea bed, hydrothermal vents and extreme marine environments. A decade later, the United Nations University – Institute of Advanced Studies (UNU-IAS) produced a publication building on many of the points proposed by Glowka. See Arico, S. and Salpin, C. (2005) *Bioprospecting of Genetic Resources in the Deep Sea Bed: Scientific, Legal and Policy Aspects*. UNA-IAS. Yokohama, Japan. Available at http://i.unu.edu/media/unu.edu/publication/28370/DeepSeabed1.pdf
5 The objective of the Nagoya Protocol in Article 1 is

> the fair and equitable sharing of the benefits arising from the utilization of genetic resources, including by appropriate access to genetic resources and by appropriate transfer of relevant technologies, taking into account all rights over those resources and to technologies, and by appropriate funding, thereby *contributing to the conservation of biological diversity and the sustainable use of its components*.
>
> (Italics added)

6 The problem of fungibility can be described as a simple swap of financing.
7 Intervention #4953. Posted on April 17, 2013, 13:37 UTC by Dr. Marco D'Alessandro, Switzerland.
8 The determination of the species accessed may occur at a later stage in the research process, once taxonomic studies have been completed. Often, this involves taking specimens for classification outside national jurisdictions. Modern taxonomy involves using DNA and bioinformatics and other tools to classify biodiversity. For a practical overview of the

taxonomic process, see the CBD sponsored "The Taxonomy Initiative" at www.cbd.int/gti/

9 References to "non-monetary" benefits are unavoidable in any ABS text. For a list of non-monetary benefits, see Appendix I of the Bonn Guidelines on Access to Genetic Resources and Benefit Sharing (2004), available at www.cbd.int/doc/publications/cbd-bonn-gdls-en.pdf

10 The notion of disclosure – as a way to link IP to ABS – was first suggested by Brendan Tobin from SPDA in the early 1990s and was resisted by countries of the Andean Community for being "strange". The first legal text to include disclosure was the Peruvian national regulation for the protection of breeders' rights. Supreme Decree 008-1996-ITINCI, the regulation on plant breeders' rights, was approved in June 1996. Article 15 of this regulation establishes that the application to obtain a breeder's certificate must contain or attach, as appropriate, "the geographical origin of the new protected varieties raw material including, if the case be, the document accrediting the legal provenance of genetic resources, issued by the National Competent Authority on access to genetic resources."

11 The concept of a "certificate of origin" – stressing legality in access – was conceptualized in the context of ABS during the 1994–1996 Andean process to develop a regional regime on ABS. The concept is included in the IUCN/SPDA Technical Report to the Andean Community (at the time the Andean Pact or Cartagena Accord) on a set of elements for a regional ABS regime. See SPDA/CDA-UICN (1994) Hacia un Marco Legal para Regular el Acceso a los Recursos Genéticos en el Pacto Andino: Posibles Elementos para una Decisión del Pacto Andino sobre Acceso a los Recursos Genéticos. Reporte Técnico Legal preparado por el Centro de Derecho Ambiental de la Unión Mundial para la Naturaleza (UICN) para la Junta del Acuerdo de Cartagena con la asistencia técnica de la Sociedad Peruana de Derecho Ambiental (SPDA). JUN/REG.ARG/I/dt.4, 31 Octubre de 1994. This report is included in Ruiz, M. (1999) *Acceso a Recursos Genéticos. Propuestas e Instrumentos Jurídicos.* Sociedad Peruana de Derecho Ambiental, Lima, Peru, pp. 7–35. Though its early background is blurred and undocumented, Brendan Tobin refers to a paper by David Downes (former Center for International Environmental Law – CIEL – lawyer) in which an international certification regime is proposed in the context of the Convention on the International Trade in Endangered Species (CITES). Tobin also mentions a prior reference to the idea by Veit Koerster, former Norway CBD negotiator in an unpublished paper. Note that in CITES the vehicle of the genetic resources is indeed important as natural information may be expunged through its loss. However, a category mistake causes what makes sense for CITES to be applied to the Nagoya Protocol. (Personal conversation with Brendan Tobin, Lima, October 13, 2014).

12 In 2004, Joshua Sarnoff, a professor at Washington College of Law and patent attorney, prepared upon request of Public Interest Intellectual Property Advisors (PIIPA) in support of the Peruvian National Biopiracy Prevention Commission a memorandum on "Compatibility with Existing International Intellectual Property Agreements of Requirements for Patent Applications to Disclose Origins of Genetic Resources and Traditional Knowledge and Evidence of Legal Access and Benefit Sharing." This is one of the most substantive analyses on the link between IP and ABS to date. In short: it concludes that disclosure in the context of ABS, whether to demonstrate legal access or simply the origin of the genetic resource is fully compatible with international IP law and TRIPS in particular. Available at www.piipa.org/index.php?option=com_content&view=article&id=91

13 The first regulation ever to include specific disclosure of origin and legal provenance was Supreme Decree 008-1996-ITINCI, the Peruvian regulation on plant breeders' rights of June 1996. Article 15 of this regulation establishes that the application to obtain a

breeder's certificate must contain or attach, as appropriate, "the geographical origin of the material incorporated and used in the new protected variety including, if the case be, the document accrediting the legal provenance of genetic resources, issued by the National Competent Authority on access to genetic resources." Decision 391 was approved by the Andean Community in July 1996. Decision 391 specifically establishes in its Second Complementary Provision that Member States (of the Andean Community) "will not recognize rights, including intellectual property rights, over genetic resources, derived or synthesized products or the intangible component, accessed or developed through an activity not in accordance with its mandates." Its Third Complementary Provision determines that national IP offices of Member States must "request the applicant the registration number of the access contract and a copy of it, as a prerequisite for the granting of the right." Decision 486, the Andean regime on IP (2000), is even more specific. Not only does it condition the granting of IP to respecting and safeguarding Member States' interests in biodiversity (Article 3), but it includes specific procedural provisions in this regard. Article 26(h) establishes that the patent application should include as appropriate "a copy of the access contract, when the product or procedure for which a patent is being requested, have been obtained or developed based on genetic resources or derived products of which Member States are countries of origin." Article 26(i) establishes that the patent application should also include as appropriate "the document which provides evidence of the license or authorization of use of traditional knowledge of indigenous communities … when the product or procedure for which a patent is being requested, have been obtained or developed based on genetic resources or derived products of which Member States are countries of origin." Disclosure extends to traditional knowledge. Article 75(g) and (h) establishes that if these requirements are not met, the patent can be annulled.

14 Attempts have been made to modify international patent law, specifically TRIPS, with no success. Efforts to streamline disclosure into IP have a long-standing history, going back to when the Committee on Trade and the Environment of the WTO and during the TRIPS review process under the Doha Round began discussing the matter. At the national level a few countries, including developed countries such as Norway and Switzerland, have already amended their IP legislation, requesting disclosure – albeit for PIC and MAT. Under the TRIPS Agreement, disclosure is mandatory. Article 29.1 of TRIPS mandates that WTO members:

> shall require that an applicant for a patent shall disclose the invention in a manner sufficiently clear and complete for the invention to be carried out by a person skilled in the art and may require the applicant to indicate the best mode for carrying out the invention known to the inventor at the filing date or, where priority is claimed, at the priority date of the application.

Still debated is whether disclosure of legal access or origin of species is essential "for the invention to be carried out by a person skilled in the art." Modifications to Article 29 of TRIPS to accommodate legal and geographical origin of genetic resources have been proposed since 2000 onwards. For a comprehensive review of the need for supportiveness between TRIPS, IP, ABS and TK, see Chouchena-Rojas, M., Ruiz, M., Vivas, D. and Winkler, S. (2005) *Disclosure Requirements: Ensuring Mutual Supportiveness between the WTO TRIPS Agreement and the CBD*. Gland and Cambridge: IUCN; Geneva: ICTSD. Available at www.ciel.org/Publications/DisclosureRequirements_Nov2005.pdf

15 Many patents and patent applications already disclose origin of species, especially so in natural product chemistry. However, the disclosure is purely voluntary. In 1995 Asha Sukhwani, of the Spanish Patent Office, published a book *Patentes Naturistas*. See Sukhwani, A. (1995) *Patentes Naturistas*. Madrid: Oficina Española de Patentes. The study on patent applications for microbial inventions by Oldham *et al.* also

demonstrates silence in patents with respect to the origin of species. See Oldham, P., Hall, S. and Forero, O. (2013) "Biological Diversity in the Patent System." *PLoS ONE* 8(11): 6.

16 In Norway and Switzerland, disclosure of MAT and PIC is voluntary and does not condition patent processing or its validity. Redress is offered through compensation. In the Andean Community, Brazil, Costa Rica and Panama, disclosure operates as a formal and substantial condition for patentability, i.e. a *de facto* patentability requirement.

17 See paper by Vogel, J.H. (2009) "iBOL as an Enabler of ABS and ABS as an Enabler of iBOL." Presentation to the Second International Conference of iBOL. Mexico City, November 7–12, 2009. In *Proceedings of the Seminar "Barcoding of Life: Society and Technology Dynamics – Global and National Perspectives"*, pp. 38–47. Submitted by the International Development Research Centre of Canada. Available at www.cbd.int/doc/meetings/abs/abswg-09/information/abswg-09-inf-15-en.pdf

18 iBOL was created in 2003 by researchers at the University of Guelph in Canada. iBOL was not created with ABS in mind. iBOL became quickly aware of the current implications of ABS for taxonomic research. See UNEP/CBD/WG-ABS/9/INF/15, March 10, 2010, *Proceedings of the Seminar "Barcoding of Life: Society and Technology Dynamics – Global and National Perspectives"*. Submitted by the International Development Research Centre of Canada. Available at www.cbd.int/doc/meetings/abs/abswg-09/information/abswg-09-inf-15-en.pdf

19 See http://en.wikipedia.org/wiki/Holdridge_life_zones

Chapter 6

Conclusions and recommendations

The CBD is a framework convention and evolves through the Decisions of the COP. The negotiation of the draft of the convention, its signature at the Earth Summit Rio '92, and entry into force in November 1993 were all premised on the convention being, precisely, framework. Nothing justifies its status as framework more than the ability to correct a foundational flaw. Yet, perversely, a wide range of stakeholders and delegates to the COPs, beginning with COP 1, have resisted correcting the misclassification of "genetic resources" as "material" (Article 2). They have not only ignored the scholarly literature but militated against correction, which leads quickly and easily to a multilateral approach.

The COP 12 held in October 2014 in Pyeongchang, Korea, which marks the entry into force of the Nagoya Protocol, follows suit and – among other of its interventions and decisions – only "invites parties and others, including ILCs, and relevant stakeholders to submit to the Secretariat views on: situations that may support the need for a GMBSM that are not covered under the bilateral approach" (Earth Negotiations Bulletin, October 20, 2014).

Alas, the foundational error of the CBD implies that almost *all* situations are not covered by the bilateral approach. Putting it another way, under a bilateral approach to ABS no regime can result in fairness and equity in the sharing of benefits for reasons reiterated throughout the book. The "views" on such "situations" have been submitted and extensively discussed in at least three different circumstances:

- The First Reflection Meeting on the Global Multilateral Benefit Sharing Mechanism (2011)
- The Expert Meeting on Article 10 of the Nagoya Protocol (2013), and especially
- The Online Discussion on Article 10 of the Nagoya Protocol (2013) – where over 140 experts participated and expressed their views.

The CBD Secretariat prepared a lengthy Synthesis of the Online Discussions which expunged the discussion of the "foundational flaw" in the definition of "genetic resources" as material (ABS Clearing House, 2013, Interventions #4871, #4891, #4912, #5028, #5284). Also expunged was any mention of the 22 references to "the economics of information". Fortunately, in the age of the internet, the memory hole is no longer unfathomable and the interventions are available on the CBD website for those rightly skeptical of the Synthesis.

Although lawyers are accustomed to argue by analogy, the strongest argument for genetic resources as natural information lies in its homology with artificial information: equal treatment through intellectual property protection. "Bounded openness" recognizes that natural information will always be open *de facto* and bounds must be placed *de jure* over the disclosure of utilization and the subsequent sharing of royalties among stewards and Providers. The share of collected royalties to a steward will reflect the geographic proportionality of the stewardship. The equity and fairness results in incentives for *in situ* conservation, which bilateral contracts do not necessarily provide. The Nagoya Protocol provides a window of opportunity to correct the foundational flaw of the CBD that has beleaguered all twelve COPs: Article 10 and Article 11 concern transboundary resources, situations where PIC is not possible, and shared genetic resources, which means all species that are not endemic.

Bounded openness flows from a synthesis of the economics of information and the economics of transaction costs. The application is sufficiently simple that it can be explained to the lay public unschooled in the economics or even the acronyms of the literature. Nevertheless, to date, the COP discussions have not drawn from the economic literature on ABS, other than acknowledging ambitious attempts to construct the total economic value (TEV) of biodiversity and genetic resources that include the hotly contested value of pharmaceutical bioprospecting.

Above all, the policy implications of the correct classification of genetic resources as natural information provide robustness for the GMBSM and enable the regime to grapple with scientific and technological scenarios largely unforeseeable when the CBD was drafted in 1992.

Curiously, participants in ABS discussions do agree that the "-omics" revolution presents challenges for policy and law. The field scientist as hero (typified by Sean Connery in the 1992 film *The Medicine Man*) seems largely irrelevant in the current technological environment. However, these same participants resist acknowledging that classification of genetic resources as natural information could resolve ABS issues both decisively and elegantly. The advent of synthetic biology is a case in point. With genetic resources misclassified as material, Users from synthetic biology will claim that they have not accessed any material and are therefore under no ABS obligation. Logically, that follows. With natural information as the object of access, all of synthetic biology falls under an ABS obligation as its business model is to secure economic rents through time limited IP. Something similar can also be said of biomimicry and the observation of non-human cultures which can inspire R&D without any access to genetic material. Natural information sweeps in all such scenarios.

Rather than correcting the foundational flaw, we see it compounding over the COPs. Time and time again, "genetic resources" as "material" is reinforced through sponsored courses, capacity building programs and Global Environment Facility (GEF) initiatives and projects. Criteria for fairness and equity are eclipsed by the intricacies of PIC, MAT, contracts, User measures, non-monetary benefits, upfront payments, and so on. However, the royalty rate goes unsaid. In other words, the only real indicator is shunted aside to be negotiated bilaterally under secrecy.

Occasionally royalty rates are leaked and no one should be surprised when they are reported at typically 1 percent or less. The economics of information predicts such an outcome.

Explanations vary as to why the ABS community does not change direction when a reservoir of scholarship exposes the foundational flaw and its consequences for the first two objectives of the convention: conservation and sustainable use. To the usual explanations of path dependency and principal-agent problems, arises a somewhat reductionist dimension. As a consequence of human evolution as a eusocial species, delegations may be eager to want to belong to the dominant group and therefore dismiss logic and evidence (Vogel 2013). They resist any attempt to question and critique. The naturalistic fallacy of ethics triumphs: what is should be. Again, this is especially bizarre given that the CBD is a framework convention and should be dynamically adaptive to changing circumstances. No one expects immediate reform. However, more than two decades have passed and the contentious issues of ABS are the same as 20 years ago and even aggravated by the schism between the advancement of science and technology and the route taken by ABS policy and regulatory frameworks.

Because the correct classification of genetic resources is undeniable, stakeholders and delegates to the COP will often equivocate and even concede that genetic resources are information yet, in the same breath, continue to treat them as if they were a tangible. People in power do not lightly admit to an egregious error much less act on the consequences of a correction suggested by critics. Nevertheless, both Users and Providers will benefit immensely from correction as bounded openness is the truly win–win situation for principals. One must recall that the lawyers and bureaucrats are merely agents. Users will be able to undertake commercial or non-commercial research, unencumbered; Providers will be able to pursue collaborations without worry of unwittingly facilitating biopiracy. Users and Providers need to agree on three core propositions:

- rents on natural information are justified for all countries which steward natural information, in proportion to their stewardship calculated on the basis of geography and distribution of species;
- royalties will be the vehicle for rents on goods and services that derive from natural information; and
- disclosure of the utilization of natural information will be woven into the securement process of IP (patents may be a good starting point for the system to operate).

A new international ABS regime based on bounded openness would generate enormous benefits as it would extend to all IP that deploys natural information, not just patents. The magnitude of the monetary benefits will be a function of the royalty rates negotiated for the type of IP and other characteristics. The transaction costs of the regime will plummet as Certificates of Compliance, PIC and MAT are not needed.

What political steps should be taken to make "bounded openness" operational? In the Foreword to this book, Vogel identified ten theoretical steps which constituted the critical mass of "bounded openness" and another five steps that strengthen the acceptance or efficiency of the solution. As Vogel showed, other economists at different institutions and points in their careers converged on the elements of "bounded openness" in the early 1990s. They also published in top-tier venues and were studiously ignored. In order that this book not suffer a similar fate, some recommendations follow for future COPs, especially COP 13 to be held in Mexico in October/November 2016.

THE AGENDA FOR REFORM

1. Address the definition of genetic resources for the purpose of utilizations which will carry intellectual property rights, whereby the current definition must compete with suggested alternatives.
2. Address the definition of "sovereignty" for the purpose of utilizations which can be conducted in one of a number of jurisdictions.
3. Address the bias in "country of origin" when the object of utilization may not be unique to one species or one species endemic to one country.
4. Address the possibility that the utilization of the object of access extends to all intellectual property and not just to patents.
5. Address the contradiction that under MTAs and ABS contracts citizens do not know the royalty rates for which their elected officials are negotiating, basically selling the genetic patrimony through granting PIC.
6. Address the implications for the GMBSM of exponential decay in the cost of determining the genome of any species and its geographic distribution (e.g. iBOL, GBIF and so on).
7. Address the understandable resistance of a megadiverse Provider and User country to ratify the CBD and why that country would be richly rewarded as a magnet for the relocation of R&D from Ratified Parties in light of the Nagoya Protocol.

If delegates from both User and Provider countries were to adopt the above agenda for reform and succeed, what would bounded openness look like? Seven features are easily imaginable:

1. Intellectual property rights legislation would be modified worldwide to require mandatory disclosure of natural information used in the utilization.

For patents and inventions, disclosure may be a simple tick in a box Yes or No. For copyrights, disclosure may be a modification of the copyright symbol to include an abbreviation for natural information; for trademarks, a modification of the trademark symbol, and so on.

2. Scientific analysis (using iBOL, GBIF, GPS, and other available technologies) would be undertaken to determine the taxon in which the natural information is found and the geographical range of organisms belonging to that taxon, once a utilization is sufficiently successful to justify the costs of such determinations.

3. As part of the GMBSM, a Global Fund would be established to receive the royalties on sales of products that utilized natural information and carried intellectual property protection, where disbursement would be calculated according to the percentage of habitats in the taxon in which the natural information is found.

4. To detect non-compliance, holders of intellectual property rights (patents, copyrights, trademarks, and so on) would be tracked and monitored through, say, patent applications from the United States Patents and Trademarks Office (USPTO), European Patent Office (EPO) and Japanese Patent Office (JPO) and similar searches for the other categories of IP.

5. Non-compliant Users would face penalties calculated to discourage non-compliance where the probability of being detected multiplied by the penalty of non-compliance is greater than the royalty due.

6. Basic research in the natural sciences will flourish as the transaction costs of access are greatly reduced albeit not eliminated as phytosanitary, CITES and other more classic restrictions, such as collecting permits and others, will continue to apply.

7. Research and development in any utilization of natural information will be better distributed among developing countries and between developed and developing countries.

The Turkish proverb that begins this book will also end it. No matter how long you have traveled down the wrong road, turn back. The reading of this book is a rest area. Looking back, the desired destination is tantalizingly close.

Case Study 1

Epipedobates anthonyi under "bounded openness"[1]

Klaus Angerer

Epibatidine illustrates the pitfalls of the bilateral approach and the advantages of bounded openness for ABS. A counterfactual history suggests that the most contentious issues surrounding the poison dart frog *Epipedobates anthonyi* would not have arisen under bounded openness.

The alkaloid epibatidine was first isolated from the secretions of *E. anthonyi*, which is endemic to south-western Ecuador and northern Peru. The discovery was deemed a decisive contribution to pharmaceutical research and "a possible first step toward producing a long-sought drug: a powerful non-sedating, non-opioid painkiller" (Bradley 1993: p. 1117). Research and development based on secretions of the thumbnail size frogs was frequently cited in the press about bioprospecting in the 1990s. In *The Future of Life*, E. O. Wilson hailed the discovery as an example of the enormous potential value of biodiversity while NGO campaigns condemned it as a flagrant example of biopiracy and an "invasion of the frog-snatchers" (Saavedra 1999).[2]

Inasmuch as the frogs were collected in 1976, any effort to comply with the ABS framework of the 1993 CBD and/or the 2010 Nagoya Protocol (which came into force only in October 2014) would first have to establish retrospectivity. Any such endeavor would be difficult but not impossible, as argued forcefully by Preston Hardison of Tulalip Natural Resources in the Online Discussion Groups on Article 10 of the Nagoya Protocol. Serious legal scholarship can be cited in support of such a stance (Sampford 2006). However, this thought experiment about bounded openness should not be confused with retrospectivity. Rather than exploring restitution for consequences of prior events that would now be illegal, one can explore what would have been the consequences of prior events had the frogs been accessed under a bilateral versus the multilateral regime of bounded openness.

Some core facts of the case must be reviewed before one can engage in a counterfactual history. Despite the media hype over epibatidine, the compound has not succeeded to date as a drug lead. Although not money-making, epibatidine has been path-breaking for opening up streams of research. As a tool, various forms of the compound are sold in bulk for prices ranging from US$20 to US$40 per mg (Sigma-Aldrich 2013; TOCRIS Bioscience 2013) and occasionally epibatidine can be found on Ebay, even being offered by apparently private sellers.[3] Inasmuch as no derivatives have yet reached the market, the monetary benefits have been very low.

To assess what would have been the value of epibatidine under bounded open-ness, one must first establish a timeline of events (for a detailed account, see Angerer 2013). In 1974, the US National Institutes of Health (NIH) sent the chemist and pharmacologist John Daly to lead an exploratory field trip to Ecuador and capture specimens of interesting poison dart frogs. The team would collect just skins. Early references misidentified the samples as *Epipedobates tricolor* but a taxonomic revision much later established that the samples belonged to *E. anthonyi* (Darst et al. 2005: p. 59). Once back at the NIH, Daly injected extracts of the skins into mice. The effect was unexpected: the mice arched their tails over their backs in a phenomenon known as the "Straub tail reaction", which is typical of opioids but hitherto unseen in frog alkaloids. What compound was responsible for the reaction? Interest became heightened when the extract was shown to have powerful analgesic properties on the mice.

Having exhausted the initial supply of skins in experimentation, Daly returned to Ecuador in 1976 to collect more samples. Hopes were high in discovering a potent opioid in the poison. However, the frogs at one of the previous collection sites had disappeared and the secretions of a nearby population contained no alkaloids. At another site, the researchers gathered skins from 800 frogs but, back in the lab, isolated no more than 500 µg of so-called "Straub tail alkaloid." By 1978, Daly had realized the far-reaching implications of his discovery. Because the effect of the alkaloid was not blocked by opioid antagonists, the analgesic could not be an opioid. Hence, the discovery could potentially eliminate the risk of dependency characteristic of opioids. Moreover, in comparison to morphine, the alkaloid was 200-fold more powerful.

With the existing technology and only minute quantities of the 1976 sample left, the structure of the compound could not be established. More frog skin extracts were required. Yet on all subsequent trips to Ecuador, the research team found specimens with only insignificant amounts of alkaloids in their secretions. Frogs raised in captivity were also free of the alkaloid (Williams et al. 2009: p. 207; Daly et al. 2000: p. 132). Hence, the samples from the original collection had become "irreplaceable" (Daly 1998: p. 169).

Further access to *E. anthonyi* became encumbered in 1987 due to the listing of the family to which the species belongs (Dendrobatidae) in Appendix II of the Convention on International Trade in Endangered Species of Wild Fauna and Flora (CITES).[4] The frogs were now subject to strict regulations that required authorization and a non-detrimental finding by the exporting country.[5] Daly writes that it had become nearly impossible "to obtain permits to collect the requisite hundred or more specimens required for structure elucidation of minor and trace components found in dendrobatid frogs" (Daly 1998: p. 169). Due to the aforementioned problem of insignificant alkaloids in the post-1976 collections and the difficulty of accessing additional specimens after 1987, Daly presciently cryopreserved the 1976 samples believing that technological advances would someday make the minute supply sufficient for elucidation of the molecular structure.

Technological advances in the 1990s did indeed revolutionize natural product chemistry. The sensitivity of NMR (nuclear magnetic resonance) spectrometers increased by orders of magnitude and the collection of more frog skins became unnecessary. Taking a calculated risk, Daly's group subjected their unique sample to NMR analysis and happily solved the molecular structure of epibatidine (Williams *et al.* 2009: p. 210; Spande *et al.* 1992). Remarkably, it was the minute sample from the 1976 collection that ultimately allowed determination of the molecular structure of the compound, some 16 years later.

One indicator of the impact of the 1992 publication of the structure are the 300+ scientific papers published in the following decade. An entire issue of the journal *Medicinal Chemistry Research* was devoted to the alkaloid (Dukat and Glennon 2003: p. 365). Furthermore, on March 3, 1992, Daly and his colleagues filed for a patent.[6] Scientific interest in epibatidine was subsequently increased by a popularized article published in *Science* (Bradley 1993). Soon afterwards, several syntheses of the alkaloid were also reported. The mechanism of analgesic activity was solved the following year: blockage by a nicotinic antagonist (Bradley 1993; Badio and Daly 1994). Notwithstanding the sweeping implications of epibatidine for opening up a new research stream, the alkaloid itself was never developed into a drug because of a narrow therapeutic window with analgesic effects "accompanied by adverse effects ... at or near the doses required for antinociceptive efficacy" (Bannon *et al.* 1998: p. 77). The contribution of epibatidine in drug discovery is mainly as a research tool in the development of synthetic derivatives. But had the skins been collected post-CBD ratification, would the synthetics qualify as derivatives and be an object for the sharing of benefits?

The answer is not clear. The trajectory of research on epibatidine highlights a feature typical of natural product chemistry: basic research within public institutions – in this case, the NIH – creating a platform for commercial research characterized by informal collaborations among public and private institutions. By the early 1990s, Abbott Laboratories had already pursued years of research on nicotinic cholinergic receptor (nAChR) agonists, which is the substance class to which epibatidine belongs (Arneric *et al.* 2007: p. 1094). Previous research on nAChR agonists could suggest that they were at the cusp of developing related substances. Although the program had one clinical candidate, no significant advance occurred until learning about epibatidine. One Abbott scientist relates how upon reading the report in *Science*, he "immediately recognized that NCEs [new chemical entities] with similar structural motifs were being made at Abbott" (Arneric *et al.* 2007: p. 1097). Another chemist from the nAChR group writes how he

> then immediately contacted Daly to see whether the MoA [mechanism of action] of epibatidine was known. John [Daly], in his usual gracious manner, indicated that a paper was in press on this topic and after being asked whether it was nicotinic receptor-mediated, agreed.

> (Arneric *et al.* 2007: p. 1097)

Before elaborating the counterfactual history, one should think like a lawyer regarding the actual history. The contact between Abbott scientists and Daly did not involve any transfer of materials. The sharing of information prior to publication was also informal without contracts or caveats of non-disclosure. Since the mechanism of action was later published and the structure already disclosed, Daly did not reveal any details that would have remained beyond the reach of Abbott's scientists through just reading the scientific literature. Nevertheless, the consultation probably did accelerate the development of derivatives and may have been instrumental in reassuring Abbott scientists that further research on nAChR-based analgesics was worthwhile. One comments: "[until] the discovery of epibatidine, it appeared that there existed an affinity barrier or ceiling beyond which it was almost impossible to venture" (Dukat and Glennon 2003: p. 365). Abbott then used the knowledge of the properties, structure and mechanism of action of the alkaloid as an inspiration for the design of a library of more than 500 optimized compounds related to the mechanism of action in epibatidine; yet it is almost impossible to precisely assess the extent of the inspiration provided by the frog alkaloid for the design of these derivatives. Screening led to the identification of ABT-594, a compound as potent as epibatidine but without the severe side effects (Williams *et al.* 2009: p. 211). An article about the drug lead was published in *Science* (Bannon *et al.* 1998) which led to "full media coverage, print, radio and cable TV" (Arneric *et al.* 2007: p. 1097).

The scientific discovery pierced the public sphere as the narrative led to a dramatic storyline. The hope for a frog-derived painkiller was even put into song in Paul Simon's "Señorita with a Necklace of Tears" (2000).[7]

Press coverage about ABT-594 sparked accusations against Daly and Abbott for having committed biopiracy. The ensuing debates focused on the question of whether the collection of the frogs in the 1970s complied with regulations then in force and whether traditional knowledge was used. Always noted was the absence of any benefit-sharing agreement between Ecuador and Abbott.

The controversy proved difficult to resolve. Opinions differed regarding the very existence of a competent authority to grant access in Ecuador in the mid-1970s and thus the legality of the export of the samples. Absence of archival records meant that the delay between the collection of the specimens and the use of their alkaloids rendered impossible any proof of the legitimacy or illegitimacy of the access to the frogs.[8] Despite such lacunae, the Instituto Ecuatoriano Forestal y de Áreas Naturales y Vida Silvestre (INEFAN) filed a claim in 1998 that Abbott should share any benefits derived from the knowledge of the indigenous communities with the Ecuadorian State.[9] Two salient facts ran counter to the claim: the frog species was probably not used locally for dart poisoning and no revenues had been generated by ABT-594. Although the claim failed (Ribadeneira 2008: p. 104), both Ecuadorian and international NGOs started a campaign accusing Abbott and Daly of biopiracy.[10]

In the proposed thought experiment on bounded openness, the sequence of events suggests several questions. Had the proposed bounded openness regime (or a new GMBSM) been binding on Abbott (or the US), would they or any other enterprise that commercializes derivatives been liable for sharing of benefits? The answer lies in the timeline: inasmuch as the discovery of the *natural* information on the frog alkaloid was first published by Daly, ABS obligations would have kicked in. However, had the same information (in this case, the same molecular structure or the same mechanism of action) been first invented by Abbott – and they were close to a similar discovery in their investigation of compounds binding to the same class of receptors as epibatidine – then there would have been no obligation. In that case, the information would have first been invented artificially rather than discovered naturally.

The thought experiment continues. Would a royalty levied for derivative uses also have to be shared with traditional communities? Although reports in reputable sources (Bradley 1993) as well as NGO briefs claimed that secretions from *E. anthonyi* were used as a dart poison and inspired Daly's endeavors, no traditional use of the species has ever been documented (Angerer 2011: pp. 360–361). Even if the claims of TK were accurate, any such usage would probably meet the criteria of public domain knowledge as long as no efforts were made to keep it secret. So, the royalty collected would not have to be shared with the communities.

Other broad lessons emerge by imagining a counterfactual history based on a system of bounded openness over natural information implemented through the proposed GMBSM. The first is "jurisdiction shopping" in transboundary situations. Although Daly and his colleagues had been gathering samples from Colombia and Panama since the 1960s, the NIH frog alkaloid program would eventually become global in reach. Live frogs and/or frog skins were collected from Argentina, Australia, Brazil, Colombia, Ecuador, Madagascar, Panama, Peru, Thailand and Venezuela and amphibian toxins were isolated from more than sixty species (Daly *et al.* 2000: p. 131; Gillis 2002). Not surprisingly, Daly's group preferred collection of species with ready access and stated so frankly:

> The research has been hindered by difficulties in obtaining permits to collect any amphibians for scientific investigation, especially in neotropical countries of Central and South America, where the alkaloid-containing dendrobatid frogs are found. For this reason, in the past decade our research has shifted to bufonid frogs of Argentina and to mantellid frogs of Madagascar.
>
> (Daly 2003: p. 449)

In other words, Daly engaged in "jurisdiction shopping". Since many species can be found in more than one country, such behavior can be expected under the bilateralism in ABS (i.e. contracts based on PIC and MAT) and may be the driver of "price wars" among countries which are "sovereign" (!) over shared resources. Each country reasons that even a pittance is better than nothing and so begins the race to the bottom.

Abbott dismissed the allegation of biopiracy and, according to Martinez-Alier, claimed that it "owes nothing to Ecuador because it merely got the inspiration for its drug by reading a scientific paper about the frog chemical" (Martinez-Alier

2002: p. 134). The dismissal suggests the inability to monitor what the Nagoya Protocol calls the "subsequent applications" of genetic resources, which are often years after the original access.[11] The relationship between first physical access to samples *in situ* and future commercial use is seldom (if ever) straightforward as various institutions, often unbeknownst to one another, become involved and technology starts to kick in. The bilateral approach does not recognize unpredictable trajectories, presupposing compliance upstream and requiring compliance downstream. In other words, it is based on trust which when violated leads to variants of the refrain "well, we were told this material was accessed in compliance with [fill in the blank.]" The epibatidine case of disembodied information puts in high relief the impossibility of such ABS obligations as the physical sample – in CBD terms, the "genetic material" – was never transferred to Abbott.

An unwelcome lesson may have been learned by the pharmaceutical industry: do not disclose any origin in natural product chemistry when the compound could have been synthesized. A dishonest assertion of convergence would have been most credible. So, if Abbott had simply patented ABT-594 without any reference to epibatidine in its publications on the compound, the bad publicity would have been averted. Morten Tvedt and Ole Fauchald highlight a similar point about enforcement of benefit-sharing in Norway:

> The first challenge the provider country faces is to become aware of the fact that a genetic resource originating in that country is being utilized in Norway, as Norway's ABS regime does not provide any guarantee that a relevant provider country will be so informed. … In general, the use of a genetic resource has no obvious external verifiable manifestations that would be controllable by either government.
>
> (Tvedt and Fauchald 2011: p. 391)

The Nagoya Protocol does impose a series of obligations on Users, who are expected to contribute – with concrete administrative or legal measures – to enable a degree of tracking and monitoring of genetic resources.[12] However, these measures are part of "bilateralism logic" under which they seek to follow resources all the way along the R&D phases and in some way ensure conformity with original PIC and MAT.

Non-disclosure is also important for the valuation of biological diversity: an incentive not to reveal the use of natural information biases downward the expected royalties. Inasmuch as epibatidine is a rather simple molecule that could have been synthesized, the possibility of royalty evasion would also exist under the GMBSM. But other blockbuster metabolites are not simple and their very complexity is an indicator of a natural origin (e.g. paclitaxel [Taxol]).

The case of *E. anthonyi* reinforces a justified sense of futility "that as soon as the genetic resources leave the country they 'are gone' [which] is often voiced in national discussions on access" (Fernández Ugalde 2007: p. 7). The public outcry in Ecuador fostered mistrust toward any kind of specimen collection. One ethnobiologist interviewed spoke of "bioparanoia" and the neologism resonates beyond

Ecuador. The 1996 Andean Community Decision 391 also stipulates rigid ABS measures, which reflects the well-publicized accusations of biopiracy.[13] Decision 391, the Philippines Executive Order 247, the Biodiversity Law 7788 in Costa Rica, Medida Provisoria 2.186-16 are "children of their time" conceived in a context of mistrust between the South and North.

The practical impossibility of reliably monitoring and tracking the flows of genetic resources leaves Provider countries with no option other than requiring

> all guarantees possible at the point of access, which is often translated into elaborate access provisions, at an increased cost ... despite the fact that the nature and amount of benefits (if any) are highly uncertain at the outset.
>
> (Fernández Ugalde 2007: p. 8)

As the transaction costs of access rise, jurisdiction shopping is encouraged and illicit access tolerated. A vicious cycle emerges as Users seek less cumbersome ways to obtain specimens which result in ever more cumbersome Provider country measures. The proposed GMBSM obviates jurisdiction shopping, illicit access and the transaction costs of establishing the competent authority. Bounded openness would have permitted Daly to collect samples without any ABS contract or MTA – maybe through a simple one-step immediate permit.

The bilateral approach, however, arguably would have impeded Daly's endeavor for a reason seldom discussed in ABS forums: scientists often employ a trial-and-error method for identifying interesting specimens – in the case of *E. anthonyi*, by literally touching and tasting the frogs in the field – not knowing precisely which species to collect before they actually start to collect (Gillis 2002; Myers *et al.* 1978: p. 339). Thus, Daly would not have been able to obtain permits under the bilateral approach inasmuch as Provider countries usually require identification of the species to be collected before any access. Swen Renner and colleagues have detailed the bureaucratic burden which becomes a welcome "daunting legal challenge" for lawyers ready to bill as "bioprospecting consultants" (Renner et al. 2012; Watanabe and Teh 2011: p. 874). The thought experiment of the bilateral approach to *E. anthonyi* is truly chilling: to the extent that the toxin was ephemeral and elusive to Abbott researchers, the steep transaction costs would have meant that humanity would have been denied a whole research stream on analgesics.

Serendipity lay at the heart of the discovery of epibatidine. Daly did not find a species that always contains bioactive alkaloids. Like most poison frogs, the toxins secreted by *E. anthonyi* are accumulated from dietary sources, leading to variable alkaloid profiles in different populations of the same species. The dietary source of epibatidine is still unidentified but could conceivably be found among arthropods like ants or mites which might sequester toxins of plant origin (Saporito *et al.* 2012: p. 164). Despite several excursions over more than a decade, some to the same sites, the researchers only detected significant amounts of epibatidine on two occasions. What is the significance of such scarcity? Vital for discovery may be local conditions that do not persist over time, even under natural conditions. In the case of epibatidine, the alkaloid reflected the available prey of specific populations of frogs at a certain moment in time. In terms of the "economics of information", Daly had accessed ephemeral natural information. The inference for ABS policy-making is strong: the transaction costs for collection should be minimized as the object of value may go extinct even though the sampled population survives.

Another subtlety typically disregarded in ABS forums is the option of long-term cryo-storage. Daly had waited for improvements in the technology out of "fear of losing the world's last supply," storing the samples at −5°C (Williams *et al.* 2009, pp. 210 and 215). The option is routinely exercised as frog skin extracts contain varying alkaloid profiles which may someday be valuable.

Such uncertainty is common in natural product chemistry. Epibatidine may be an unusual case but it is by no means exceptional. The exact biochemical contents of genetic resources accessed for R&D are usually not known *a priori* and even less so are the revenues that someday may be generated. In contrast, technologies for long-term storage and future analysis render the preserved samples potentially useful in perpetuity. The asymmetry raises a conundrum for the bilateral approach: How to negotiate the contractual conditions for ABS based on values which may materialize several decades later with technologies presently unimaginable. Or worse, how to estimate the commercial value of the genetic resources of, say, Ecuador far into the future. The proposed GMBSM under bounded openness commits no vaulting ambition and only requires payment of royalties in cases where revenues eventuate. The ensuing royalties are then shared among the countries which could have provided the resource in question, in this case, only Ecuador because one does not know whether the alkaloid ever existed in the diet of other populations of the species. One only knows that it did exist in significant quantities at two moments in time, 1974 and 1976, and in two populations in Ecuador that happened to have been sampled by Daly as the researchers themselves highlight: "whatever the dietary source of epibatidine, it was neither abundant nor widely distributed" (Daly *et al.* 2000: p. 132).

Inasmuch as one cannot predict the potential use of frog alkaloids, one can also not predict the market value. Prior to the reports on epibatidine, little expectations existed of an economic value of poison frogs; after the reports, high expectations emerged based on misperceptions. To date, the expectations have not been fulfilled as no epibatidine derivatives have made it to market. Several related compounds are in development – for example, Abbott's compound ABT-894 (Arneric *et al.* 2007: p. 1094) – but whether such compounds qualify as "derivatives" or "subsequent applications" according to the provisions of the Nagoya Protocol is unclear. Should such compounds one day become commercially successful, the question becomes: What degree of modification renders a compound so distant from the original structure as to no longer be classified as a derivative? Despite the "hope for the development of a 'next generation' of drugs based on the epibatidine pharmacophore" (Jones *et al.* 2006: p. 257), no direct benefits have been generated to date which could be shared. With that said, one must quickly add that epibatidine has inspired research on nicotinic analgesics and drug leads for targets that had been previously unexplored. The impact of epibatidine in nicotinic receptor research and its role in ongoing research of neuronal nicotinic acetylcholine receptor ligands are highlighted in scientific overviews (Dukat and Glennon 2003: pp. 374–375; and Nirogi *et al.* 2013: pp. 23–26).

The benefits from epibatidine have been indirect and non-monetary in the realm of contributions to biological, pharmaceutical and chemical knowledge which might have otherwise been missed, as scientists involved in the development of ABT-594 readily admit: "Abbott researchers along with others in both academia and industry benefited significantly from the basic research efforts funded by the NIH" (Arneric et al. 2007: p. 1097). Expectations concerning the value of all amphibian secretions have increased enormously even though the success of epibatidine could not be replicated. Some interesting new alkaloids like phantasmidine – from the 1976 sample that also contained epibatidine – have also been identified. However, no drug leads derived from frog alkaloids have been disclosed other than epibatidine, even though research in the field has intensified as evidenced by more than 800 other alkaloids described by Daly's group (Fitch *et al.* 2010; Daly *et al.* 2005). Such uncertain prospects of generating monetary benefits complicate perceptions of the value of genetic resources and are difficult to handle within the current ABS regime based on bilateral contracts negotiated before access. Stakeholders on both sides of a contract or MTA would have to form an expectation of value when near uncertainty obtains.

A regime of bounded openness solves the problem of inherent uncertainty. Royalties would be shared only when a patented product derived from natural information generates significant revenues. An overarching implication of the poison frog for the GMBSM is the issue of flat versus tiered royalties. The case of epibatidine supports a royalty schedule with lower percentages for natural information which opens research streams where there may be hundreds or even thousands of future patented products. The rationale is similar to the attenuated compensation of standard-essential patents: not to reward the element of luck.

The question of definitions resurfaces: the object "accessed" or "utilized" in drug discovery is often just the information on the structure or properties of the compounds isolated from the sample material. Such information may be revealed through publications, databases or even casual conversations among scientists. The communication of information does not necessarily involve any transfer of materials with the correspondent formal agreements. Only in early stages of research are genetic resources researched in a purely physical medium such as extracts, fractions of extracts and isolated biochemical compounds. As a recent review of natural product chemistry stresses,

> the majority of pure NPs [natural products] represent rare chemicals of extremely limited supply. Frequently, particularly in the case of newly reported structures, such compounds are also unique commodities and are only immediately available from a single source, namely, the original investigators, or by re-isolation.
>
> (Pauli *et al.* 2012: p. 1244)

As research on compounds proceeds, value increases as the compounds become "enriched" with information on their properties or structures and constitute "informed materials" (Barry 2005). As in the case of epibatidine, the information with which these materials become enriched may be sufficient for a drug lead by someone other than the collector of the material sample. Genetic resources often become obsolete in their material form as research companies compile knowledge of their properties and libraries of isolated or derived compounds. Whenever cost-effective, derivatives are synthesized to avoid an insecure supply of the natural products. A GMBSM based on bounded openness and disclosure of natural information in patent applications greatly facilitates access in the early stages of bioprospecting or R&D while providing a simple mechanism for monitoring utilization and tracking commercialization.

One infers from this counterfactual history that the proposed GMBSM would lower precipitously the associated transactions costs in bioprospecting and thereby benefit industry. At the same time, a GMBSM could track the ultimate commercialization of natural information whenever the product enjoys time-limited monopoly intellectual property protection and ease the burden of monitoring utilization by establishing patent applications with mandatory disclosure of use of natural information as the single checkpoint. This would allow for a determination of the geographic distribution of a species and the ensuing calculation of monies to be shared in case of commercially successful inventions based on natural information.

CONCLUSIONS

This brief review of the events which led to the discovery of epibatidine and the development of derivatives suggests that in most cases the proposed GMBSM would be more adequate for the task of enabling the access to genetic resources and making sure that the potentially arising benefits are shared in a fair and equitable way. The GMBSM would presumably do justice to both users and providers of genetic resources by facilitating access as well as making sure that potentially arising benefits are shared. Thus, it would reverse the burden of ABS and aim at facilitating an open access to genetic resources particularly in the early stages of research while providing mechanisms for a tightened monitoring of their utilization in later stages of drug discovery and development where benefits are much more likely to be generated. A system of bounded openness would also avoid the fallacy of anticipating the value of genetic resources and help to lower the often substantial transactions costs at the point of access – a considerable advantage given the insecure prospects and low profitability of many cases of bioprospecting. At the same time, a GMBSM could provide mechanisms for tracking the ultimate commercialization of natural information derived from genetic resources in order to enforce the payment of royalties in cases where products enjoy monopoly intellectual property protection.

NOTES

1 As highlighted in Chapter 5, we understand the GMBSM as more than a simple fund or benefit distribution mechanism. We interpret it to mean a broader mechanism – a *regime* so to speak, with its own rules, principles and institutional set-up based on bounded openness.

2 See Wilson, E.O. (2002) *The Future of Life*, 121–123 New York: Random House.

3 See examples at www.sigmaaldrich.com/catalog/product/sigma/e0280?lang=en& region=PE and www.ebay.com/itm/Epibatidine-dihydrochloride-hydrate-poison-tree-dart-frog-secretios-/251011027148?pt=LH_DefaultDomain_0&hash=item3a716c50cc (accessed October 20, 2014).

4 CITES is an international convention adopted in 1973 which regulates international trade in endangered species (animal and plant specimens, samples, skins, etc.), through the use of an export and import permitting system (based on the level of risk facing different species) which enables international monitoring of trade (for details see www.cites.org).

5 Article IV.2 of CITES establishes that:

> The export of any specimen of a species included in Appendix II shall require the prior grant and presentation of an export permit. An export permit shall only be granted when the following conditions have been met:
> (a) A Scientific Authority of the State of export has advised that such export will not be detrimental to the survival of that species;
> (b) A Management Authority of the State of export is satisfied that the specimen was not obtained in contravention of the laws of that State for the protection of fauna and flora.

6 Patent application for "Epibatidine and derivatives, compositions and methods of treating pain" (US Patent #5314899).

7 Paul Simon, "Señorita with a Necklace of Tears," © 2000 Words and Music by Paul Simon; see www.paulsimon.com/us/music/youre-one/se%C3%B1orita-necklace-tears (accessed March 7, 2013).

8 For a detailed review, see Angerer, K. (2013) "'There is a Frog in South America/Whose Venom is a Cure': Poison Alkaloids and Drug Discovery," in Von Schwerin, A., Stoff, H., Wahrig, B. (eds.) *Biologics, A History of Agents Made From Living Organisms in the Twentieth Century*, 173–191. London: Pickering & Chatto.

9 By 1996, Decision 391 on the common Andean regime on ABS was already in force. The First Transitory Provision of Decision 391 establishes that:

> Upon entry into force of this Decision, whoever possesses with a view to access, genetic resources of which Member States are countries of origin, or derived products or intangibles associated components [TK], will have to request for the said access to the National Competent Authority, in accordance with the provisions in this Decision.

This means that any access predating the entry into force of Decision 391 needs to be regularized.

10 During the 1990s and from 2000 onwards, considerable attention was given to the case of the "Poison Dart Frog of Ecuador" as it was often referred to. Apart from the usual "biopiracy" claims by RAFI (ETC Group), complete articles and academic papers were devoted to the case. See, for example, Tidwell, J. (2002) "Raiders of the Forest Cures." *Zoogoer* (Sept/Oct): 14–21 Available at http://static.squarespace.com/static/5244b0aee4b045a38d48f8b0/t/5339967ce4b041f3867ab786/1396283004015/Raiders%20of%20the%20Forest%20Cures.pdf

11 Article 5(1) of the Nagoya Protocol determines that (among others):

1. In accordance with Article 15, paragraphs 3 and 7 of the Convention, benefits arising from the utilization of genetic resources as well as subsequent applications and commercialization shall be shared in a fair and equitable way with the Party providing such resources that is the country of origin of such resources or a Party that has acquired the genetic resources in accordance with the Convention. Such sharing shall be upon mutually agreed terms.

12 Article 17 of the Nagoya Protocol (Monitoring the Use of Genetic Resources) calls for the adoption of a wide range of measures, particularly by Users, to support the process of "following" the path of these resources along the R&D chain. The broad range of options include: designation of checkpoints; "international recognized certificates of compliance"; reporting requirements (subject to MAT); use of the ABS Clearing House Mechanism, among others.

13 See, for example, Hammond, E. (2013) *Biopiracy Watch: A Compilation of Some Recent Cases (Vol. 1)*. Penang, Malaysia: Third World Network; and Stenton, G. (2003) "Biopiracy within the Pharmaceutical Industry: A Stark Illustration of Just how Abusive, Manipulative and Perverse the Patenting Process can be Towards Countries of the South." *Hertfordshire Law Journal* 1(2): 30–47. Available at www.herts.ac.uk/__data/assets/pdf_file/0008/38627/HLJ_V1I2_Stenton.pdf

Case Study 2

Lepidium meyenii under "bounded openness"

Omar Oduardo-Sierra

Objections over the apparent misappropriation of maca (*Lepidium meyenii*) were no less vigorous in Peru than were the objections over the poison dart frog (*Epipedobates anthonyi*) in Ecuador. Due to deep historical ties to the iconic plant, any policy adopted would invariably generate controversy. Consequently, public policy appears to gauge public reaction, internalize criticisms and evolve. Albeit somewhat unforeseeably, the process suggests a willingness to attempt a profoundly different multilateral approach.

On May 9, 2003, the Delegation of Peru presented a report to the Intergovernmental Committee on Intellectual Property and Genetic Resources, Traditional Knowledge and Folklore of the World Intellectual Property Organization (IGC) on a series of questionable patents granted over Andean maca.[1] The report serves as a reference to ask the same question of maca as was asked of the poison dart frog. How would ABS have proceeded and responded under the bounded openness of the proposed GMBSM?

The introduction to the report lays the groundwork for imagining a counter-factual history:

> The patents referring to *Lepidium meyenii* or Maca are one more example, among many which exist, of how the intellectual property system – by means of patents – is based, mainly in the United States, on the privatization of biological and genetic components and materials in isolation, as part of larger inventions. In this case, these are resources in relation to which Peru (as the country of origin) has a series of rights which are not taken into account or respected. This same case refers to knowledge which, although difficult to document, has been broadly used by old Peruvians for a long period of time. This is obvious owing to the fact that many food-related, nutritional and medicinal uses of Maca, claimed in these patents, have traditionally been used by the indigenous peoples of Peru.

To apply the economics of information to maca, one separates the "biological and genetic components and material" into distinct attributes some of which are worthwhile to claim. Although maca is foremost a food crop, it is nevertheless expressly excluded from the list of 64 crops covered under the Multilateral System in Annex 1 of the ITPGRFA. Being the sole exclusion specified in Annex 1 is in itself indicative of the cultural significance of the crop for Peruvian Andean farmers. Hence, all attributes of maca would fall under the ABS provisions and principles of the CBD and the regional and national ABD framework.[2]

Nevertheless, the pursuit of intellectual property rights over all the attributes is not necessarily worthwhile. Traditional small-scale agriculture is competitive and the benefits to be shared of establishing and enforcing, say, a plant breeder's right may fall below the associated transaction costs. So, the economist would likely prioritize and focus on the "medicinal" attributes of maca given the high profitability of the pharmaceutical sector. A quick search of the USPTO reveals what patents have been issued over its various medicinal attributes (Table CS2.1).

The potential of such patents is reflected in medicinal attributes cited in the report:

> Maca is also known as an aphrodisiac which cures frigidity in women and is a remedy for impotence in men (Pulgar 1978; … García and Chirinos 1999). A great deal of evidence on successful treatment with Maca for cases of frigidity, impotence, and sterility is to be found in a Folklore Clinic in Junín (León 1986). This property of Maca could be due to the presence of prostagladins and esterols in the hypocotyl-root, and of amides of polyunsaturated fatty acids.
>
> (Li *et al.* 2001: p. 10)

The thought experiment on maca raises an issue of jurisdiction for the movement of the 64 crops listed in Annex 1 of the ITPGRFA. When one moves the material, one moves all attributes of the material. Most crops in Annex 1 will have attributes beyond nutrition, as is the case with maca. Under bounded openness of the proposed GMBSM, the object of access is not the material but the natural information bundled in the vehicle of a biological sample – listed in Annex 1 of the ITPGRFA. Article 2 (Use of Terms) of ITPGRFA establishes that "'[p]lant genetic

Table CS2.1 Patents on the value added to *Lepidium meyenii*

United States Patent and Trademark Office (USPTO) Patent Number – Brief Description	Year
6093421 – Maca and antler for augmenting testosterone levels	2000
6267995 – Extract of *Lepidium meyenii* roots for pharmaceutical applications	2001
6428824 – Treatment of sexual dysfunction with an extract of *Lepidium meyenii* roots	2002
6552206 – Compositions and methods for pharmaceutical preparations from lepidium	2003
6878731 – Imidazole alkaloids from *Lepidium meyenii* and methods of usage	2005
7985434 – Compositions of atomized or lyophilized maca (*Lepidium meyenii*) extracts	2011

resources for food and agriculture' means any genetic material of plant origin of actual or potential value for food and agriculture. 'Genetic material' means any material of plant origin, including reproductive and vegetative propagating material, containing functional units of heredity." So, the bilateral approach – based on defining genetic resources as "material" and expressed in national ABS frameworks based on PIC and MAT (contracts) which will govern their access and use[3] – would allow legal movement of every non-nutritional attribute to non-ratified jurisdictions.

The next section of the Peruvian report to WIPO also lends itself well to the thought experiment: "Brief Description of the Context: Patents, Biological Diversity and 'Biopiracy'." The quotation marks around the word biopiracy are judicious as the report proposes that the term be "understood as a political rather than a legal concept. Biopiracy refers to situations involving direct or indirect appropriation of biological or genetic resources or traditional knowledge by third parties" (p. 4). The concept invites scrutiny no matter how it is understood. Professor Charles R. McManis identifies a fallacy of equivocation in the current accusation of biopiracy as

> [O]ne cannot obtain patents or plant breeders' rights covering the knowledge or genetic resources of farming and indigenous communities, as such, but only on inventions or plant varieties that may have been derived from such knowledge or genetic resources.
>
> (McManis 2004: pp. 448–449)

Another distinguished law scholar, Professor James Ming Chen, is more pugnacious and has authored "There's No Such Thing as Biopiracy... And it's a Good Thing Too" (Chen 2006). The legal or political remedy to claims of misappropriation is rather obvious. Should patents be granted on genetic resources, then those patents should be challenged as non-statutory subject matter. This conclusion is invariant to the bilateral or multilateral approach.

Three excerpts from the report are opportunities for further exploration of the GMBSM under bounded openness:

> (1) The plant, known in Quechua as Maca, Maka, Maino, Ayak Chichita, Ayak Willku; in Spanish as Maca; and in English as Maca or Peruvian ginseng, comes from the Central highlands of the Peruvian Andes, where it has been cultivated for many centuries for its swollen roots which are edible. It is a magnificent example of a plant domesticated by the ancient Peruvians, which has helped to feed the inhabitants of Chinchaisuyo, in an environment with low temperatures and strong winds. In those areas, these climatic factors limit the cultivation of other species. For centuries Maca was used to barter for other foodstuffs or to pay taxes. (p. 6)
>
> (2) Maca is briefly described in part 1 of the work by Pedro Cieza de León, in 1553, entitled "La Crónica General del Perú" (The General Chronicle of

Peru). Vásquez de Espinoza, who visited Peru in 1598, also provides a short description of Maca in his "Compendium and Description of the West Indies," and Father Bernabé Cobo, who visited Peru between 1603 and 1629, also includes it in his "History of the New World" (Ochoa and Ugent 2001) (p. 7). (3) Although little information exists regarding the *Lepidium* species endemic to the Andes, those which are known are classified in the Dileptium and Monoplaca sections. All these, including Maca, grow in high altitude habitats, up to 4,500 meters above sea level. Brako and Zarucchi reported six other *Lepidium* species in Peru, distributed between the Departments of Ancash and Puno (Brako and Zarucchi 1993). However, some of those species are also to be found in Ecuador, Bolivia and Argentina (p. 8).

From the first excerpt from the report, one may erroneously believe that the sole country of origin is Peru. By the third excerpt, other countries of origin for the genus are identified, Argentina, Bolivia and Ecuador. Are the attributes of *Lepidium meyenii* that hold the highest expectation for a commercially successful patent found at the level of the species or the genus? Conceivably medicinal attributes could have evolved in the periphery (Argentina, Bolivia and Ecuador) and flowed into the Vavilov Center (Peru) through the sharing of seed in trade routes over thousands of years. The anthropologist Stephen Brush emphasizes such a scenario in his criticism of the bilateral approach to ABS (Brush 2010: p. 58).

What is the implication of diffusion across taxa and political boundaries? Should attributes be common to cultivars from Argentina, Bolivia, Ecuador and Peru – the last two of which were in armed conflict within recent memory – a metaphorical price war may also emerge. One can easily imagine that bioprospectors will pit one country against another, reminiscent of the quote from John Daly in Case Study 1 about shifting his research from dendrobatid frogs of the Americas to mantellid frogs of Madagascar simply because the latter were accessible, bureaucratically speaking. If the competent authority in Peru proves troublesome, then Ecuador, if Ecuador is similar, then Bolivia, if Bolivia is similar, then Argentina, and if all of the above prove troublesome, well, the extracts are already exported into the US where the CBD was never ratified. Jurisdiction shopping by the US patent applicant leads home.

The strategy of jurisdiction shopping is reinforced when uncertainty exists regarding the identity of the competent authority for access. Ironically, the case of Peru is illustrative – albeit maybe an extreme example of how *not* to design a competent authority on ABS. The national regulation to Decision 391 on ABS was only enacted in 2009, almost 13 years after the Decision came into force! But apart from the passage of time with no explicit national competent ABS authority, when the regulation did come into force, there were three different ABS administrative authorities, one for each type of genetic resource depending on its nature – wild, cultivated or domesticated, or marine.[4]

Sourcing the natural information of maca through extracts sold in US jurisdiction will frustrate any remedy sought through the ABS obligations of the CBD. Nevertheless, the implication of the largest User being a non-Party seems to elude much analysis on biopiracy. In a cited publication from almost a decade back, in regard to USPTO Patent 6267995 – Extract of *Lepidium meyenii* roots for pharmaceutical applications, it indicates that:

> [T]he maca roots that were used for these inventions were taken from Peru and there is no evidence that the material was obtained legally or that a benefit sharing arrangement was agreed between patent holders and the Peruvian state and indigenous communities. The delivery of these patents therefore runs counter to one of the three main objectives of the Convention on Biological Diversity (CBD), which is the "fair and equitable sharing of the benefits arising out of the utilisation of genetic resources".
>
> (Chouchena-Rojas *et al.* 2005: p. 30)

To the English ear, the word "taken" implies unauthorized access whereas US scientists may have simply bought the roots online (see Table CS2.2). Even if the genetic material had been spirited out of Peru or Argentina or Bolivia or Ecuador, then the landmark Supreme Court of California decision Moore v. Regents of the University of California (1990) means that the disembodied natural information would become *res nullius* once in US jurisdiction. The counterfactual history is now in stark relief. The proposed GMBSM would mitigate the forgone benefits in such scenarios inasmuch as a royalty could be levied on all US exports of the patented product into the 193 Ratified Parties.

Finally, the second excerpt from the report inadvertently shows that much TK about maca has been in the public domain, literally, for centuries. Thinking like an economist, TK is artificial information, which can be decomposed into attributes similar to natural information. A subset of those attributes may never have been published or perhaps were only invented recently and are not truly traditional. Under the GMBSM one could establish a mechanism to protect such unpublished information as trade secrets. Nevertheless, the approach would involve significant operational costs. At the level of the community, several endeavors would be necessary:

1. classification of the species associated with TK;
2. description of each attribute of the classified species and entry into a database;
3. filtering of the data uploaded against the published literature to determine what has already fallen into the public domain;
4. filtering of the non-public domain data with a similar set from the databases of other communities;
5. marketing of the trade secrets (only seemingly contradictory);
6. identification of public good projects, preferably non-fungible, to be financed from the flow of royalties to the communities of origin.

The details and software have been worked out in the anthology *The Biodiversity Cartel: Transforming Traditional Knowledge into Trade Secrets* (Vogel 2000). The transaction costs of the community databases could be borne by royalties on species bioprospected from TK in the public domain. The justification parallels the financing of iBOL through royalties collected on patents arising from R&D on ubiquitous genetic resources.

Many communities may find sacrilegious the proposal of transforming non-published TK into trade secrets. Communities which reject IP and its reification (West 2012) could decline the financing of the databases and exclude themselves from the network necessary to identify the "communities of origin". The exercise of such an option addresses the concern expressed in the Online Discussion Groups by Preston Hardison of the Tulalip Natural Resources:

> Treating traditional knowledge separately and abstractly as separate issue and applying "freedom of ideas" and intellectual property - and particularly copyright-like rules, puts their cultures at risk of harm through impacts on their customary laws on knowledge and exploitation of cultural resources necessary to their integrity, dignity, identity, self-determination, livelihoods and survival [#5316].

The aforementioned emphasis on medicinal attributes in ABS makes sense when one starts with the most profitable sectors and then moves down the list. However, one does not get very far should energy be expended in confronting the foundational flaw of the CBD misclassification of genetic resources as material. Maca provides the lesson that one should no longer wait.

Elaboration of attributes of maca, other than the medicinal, can begin with the last sectors identified in Table 5.2 of the TEEB Report, namely, "personal care,

Table CS2.2 Maca in the personal care, botanical and the food and beverage industries

Product	Price (US$)/lb	Retailer
Wild Peruvian Maca Root Powder Wildcrafted Raw Superfood	12.95	Amazon.com
Organic Maca Root Powder	13.57	Plant Spirit
Organic Gelatinized Maca Root Powder	23.68	Plant Spirit
Organic Gelatinized Maca Root Vegecaps	94.08	Plant Spirit
Organic Red Maca Root Powder Certified (raw)	19.00	Plant Spirit
Organic Black Maca Root Powder Certified (raw)	22.63	Plant Spirit
Organic Dry Maca Root Extract 5:1 Concentrate	110.00	Plant Spirit
Raw Root	60.00	Vivasos
Men's maca root aftershave (oil)	480.00	The Body Shop
Kuka Golden Ale	7.99	Andean Brewing Company

Source: Corporate websites of retailers on September 3, 2013.

botanical, and the food and beverage industries." Much variance exists within that category as easily evidenced by a perusal of prices in Table CS2.2.

In "personal care, botanical, and the food and beverage industries," what is the value added to maca which enjoys a monopoly IP? The answer varies. Perhaps maca-as-nutraceutical enhances the trademark of the intermediary (e.g. PlantSpirits or The Body Shop) but the effect is diluted by the other products advertised by the same intermediaries. Inasmuch as maca is being sold as a tangible powder by Amazon.com, economic theory would not justify any rent for ABS. In other instances, maca is strongly associated with the IP of the product. Perhaps the best example is the last item in Table CS2.2: the Kuka line of beers from the Andean Brewing Company. The website explains the meaning of the trademark:

> In native Andean language, Aymaran, 'kuka' means 'food for workers and travelers,' a reference to the bounty provided by gods. It was used to refer to the Coca plant and how its leaves played a major role in religious and social ceremonies of Andean indigenous civilizations. Maca is a main source of food and one of the very few plants that can be cultivated in the harsh climate of the Andes. And for thousands of years, Maca has provided the natives with the means to survive for generations. To the natives, Maca provides the quintessential reason to celebrate Pachamama's (Mother Earth's) bountiful gifts.
>
> (Andean Brewing Company 2013)

The company is a small-scale brewery based in New York State which competes with other small-scale breweries worldwide. The trademark is indicative of the economic theory of "monopolistic competition", pioneered by Joan Robinson. Although rents exist from product differentiation – the monopoly in "monopolistic competition" – the rents tend to be small, the "competition" partial. ABS issues are raised in at least five intellectual property rights associated with Kuka beer:

1. trade secret
2. trademark
3. geographic indication/appellation of origins
4. industrial design
5. copyright.

The label is reproduced under the fair use doctrine. Even casual inspection reveals a multitude of "traditional cultural expressions" for which no international *sui generis* intellectual property instrument presently exists (WIPO 2013, www.wipo.int/tk/en/folklore/). Looking at the label, one can make the argument that the company derives benefits from monopoly rents generated from the trademark and copyright and that the company name and industrial design pass off the geographic indication and appellation of origin. The greatest intellectual property derived from the utilization of maca is also the most subtle: the trade secrets of brewing.

Under a bilateral approach to ABS, little can be done to enforce ABS obligations of a company in a non-Party that legally imports the powder. Under the proposed GMBSM of bounded openness, a royalty could be collected on the export revenues of the derived products according to its commercial sector, the degree of substitution of the input in R&D, the type(s) of intellectual property protection and whether there is direct or indirect use. For Kuka beer that royalty would probably be a very low percentage.

The case of maca illustrates that the GMBSM under bounded openness must be nuanced and think beyond just the monopoly rents of patents.

NOTES

1 This report was the result of a national multidisciplinary working group led by the patent office (INDECOPI – the National Institute for the Protection of Intellectual Property and Competition). The group was at the time responding to claims from the NGO RAFI and national actors regarding patents over Andean maca. This group is the antecedent of the National Biopiracy Prevention Commission created in 2004 through Law 28216 enacted by the Peruvian Congress – and which is active to this date and with important successes in combating illegal and wrongful patents (see www.biopirateria.gob.pe). For the official report sent to WIPO, see Documento WIPO/GRTKF/IC/5/13. *Patentes referidas a Lepidium meyenii: Respuestas del Perú.* Quinta Sesión del Comité Intergubernamental sobre Propiedad Intelectual y Recursos Genéticos, Conocimientos Tradicionales y Folclor. 7–15 de julio de 2003. Ginebra, Suiza. This was followed by another report two years later, Documento WIPO/GRTKF/IC/8/12. *El Sistema de Patentes y la Lucha contra la Biopiratería: La Experiencia del Perú.* Octava Sesión del Comité Intergubernamental sobre Propiedad Intelectual y Recursos Genéticos, Conocimientos Tradicionales y Folclor. 6–10 de junio de 2005. Ginebra, Suiza.

2 Access and benefit-sharing rules would be covered by Andean Community Decision 391 and the national regulation on ABS (Supreme Decree 003-2009-MINAM, the national regulation for Decision 391, of 2009). Available at www.minam.gob.pe/wp-content/uploads/2013/09/ds_003-2009-minam-y-anexo.pdf

3 Article 12.3.a of the ITPGRFA establishes that access to PGRFA included in Annex 1 of the Multilateral System will be facilitated

> solely for the purpose of utilization and conservation for research, breeding and training for food and agriculture, *provided that such purpose does not include chemical, pharmaceutical and/or other non-food/feed industrial uses.* In the case of multiple-use crops (food and non-food), their importance for food security should be the determinant for their inclusion in the Multilateral System and availability for facilitated access.
>
> (Italics added)

4 The National Institute for Agricultural Innovation (INIA) has the competence/jurisdiction over cultivated and domesticated genetic resources. The Vice Ministry for Fisheries is competent over marine genetic resources and the Forest Service has competence and jurisdiction over wild plant and marine genetic resources, including microorganisms (Supreme Decree 003-2009-MINAM, Article 15) What is even more striking is that under the current review of the regulation (in its third year or so of review), there are plans to include the Natural Protected Areas Service as yet a fourth potentially competent authority in the case of genetic resources found in protected areas.

Bibliography

Andean Brewing Company (2013) *What Does KUKA Mean?* Available at http://kukablog. com/?cat=128

Andersen, R. (2008) *Governing Agrobiodiversity: Plant Genetics and Developing Countries.* Aldershot: Ashgate.

Andersen, R. and Winge, T. (2012) *The Access and Benefit Sharing Agreement on Teff Genetic Resources.* Fridjt Nansen Institute, ABS Capacity Development Initiative for Africa. FNI report 6/2012. Available at www.fni.no/doc&pdf/FNI-R0612.pdf

Angerer, K. (2011) "Frog Tales – On Poison Dart Frogs, Epibatidine, and the Sharing of Biodiversity Innovation." *European Journal of Social Science Research* 24(3): 353–369.

Angerer, K. (2013) "'There is a Frog in South America/Whose Venom is a Cure': Poison Alkaloids and Drug Discovery," in Von Schwerin, A., Stoff, H., Wahrig, B. (eds.) *Biologics, A History of Agents Made From Living Organisms in the Twentieth Century,* 173–191. London: Pickering & Chatto.

Arico, S. and Salpin, C. (2005) *Bioprospecting of Genetic Resources in the Deep Sea Bed: Scientific, Legal and Policy Aspects.* UNA-IAS. Yokohama, Japan. Available at http://i.unu.edu/media/ unu.edu/publication/28370/DeepSeabed1.pdf

Arneric, S.P., Holladay, M. and Williams, M. (2007) "Neuronal Nicotinic Receptors: A Perspective on Two Decades of Drug Discovery Research." *Biochemical Pharmacology* 74(8): 1092–1101.

Aylward, B.A. (1993) "The Economic Value of Pharmaceutical Prospecting and its Role in Biodiversity Conservation." *LEEC paper DP* 93-03. London: London Environmental Economics Centre.

Badio, B. and Daly, J.W. (1994) "Epibatidine, a Potent Analgesic and Nicotinic Agonist." *Molecular Pharmacology* 45(4): 563–569.

Bannon, A.W., Decker, M.W., Holladay, M.W., Curzon, P., Donnelly-Roberts, D., Puttfarcken, P.S., Bitner, R.S., Diaz, A., Dickenson, A.H., Porsolt, R.D., Williams, M. and Arneric, S.P. (1998) "Broad-spectrum, Non-opioid Analgesic Activity by Selective Modulation of Neuronal Nicotinic Acetylcholine Receptors." *Science* 279(5347): 77–81.

Barry, A. (2005) "Pharmaceutical Matters: The Invention of Informed Materials." *Theory, Culture & Society* 22(1): 51–69.

Berne Declaration, Brot, ECOROPA, TEBTEBBA and TWN (2013) *Nagoya Protocol on Access to Genetic Resources and the Fair and Equitable Sharing of Benefits from their Utilization. Background and Analysis.* Penang, Malaysia. p. 76.

Biber-Klemm, S., Martinez, S.I., Jacob, A. and Jevtic, A. (2010) *Agreement on Access and Benefit Sharing for Non Commercial Research. Sector Specific Approach Containing Model Clauses.* SCNAT, Bern, Switzerland. Available at www.bfn.de/fileadmin/ABS/documents/6C33Ed01__2_.pdf

Biotrade Initiative (2000) *UNCTAD Biotrade: Some Considerations on Access, Benefit Sharing and Traditional Knowledge.* Working Paper. Prepared for the UNCTAD Expert Meeting on Systems and National Experiences for Protecting Traditional Knowledge, Innovations and Practices. Geneva, October 30, 2000. Available at www.biotrade.org/ResourcesPublications/Some%20considerations%20on%20ABS%20and%20TK.pdf

Boyle, J. (2003) "The Second Enclosure Movement and the Construction of the Public Domain." *Law and Contemporary Problems* 66:33–74.

Bradley, D. (1993) "Frog Venom Cocktail Yields a One-Handed Painkiller." *Science* 261(5125): 1117.

Brako, L. and Zarucchi, J.L. (1993) "Catalogue of the Flowering Plants and Gymnosperms of Peru." *Monograph in Systematic Botany from the Missouri Botanic Garden* 45: 1–1286.

Brockway, L. (1979) "Science and Colonial Expansion: The Role of the British Royal Botanical Gardens." *Interdisciplinary Anthropology* 6(3): 449–465. Available at www.jstor.org/stable/643776

Brush, S. (1996) "Is Common Heritage Outmoded?" in Brush, S., Stabinsky, D. (eds.) *Valuing Local Knowledge: Indigenous People and Intellectual Property Rights*, 143–164. Washington DC: Island Press.

Brush, S. (2010) "The Anti-Commons Threat to Farmers' Rights: The Case of Crop Germplasm," in Vogel, J.H. (ed.) *The Museum of Bioprospecting, Intellectual Property and the Public Domain*, 55–73. New York: Anthem Press.

Burhenne-Guilmin, F. and Casey-Lefkowitz, S. (1994) "Introduction to The Guide to the Convention on Biological Diversity," in Glowka, L., Burhenne-Guilmin, F., Synge, H. (eds.) *A Guide to the Convention on Biological Diversity*. Gland and Cambridge: IUCN.

Cabrera, J. (2009) "The Role of the National Biodiversity Institute in the Use of Biodiversity for Sustainable Development: Forming Bioprospecting Partnerships," in Kamau, E.C., Winter, G. (eds.) *Genetic Resources, Traditional Knowledge and the Law: Solutions for Benefit Sharing*, 244–269. London and Sterling, VA: Earthscan.

Cabrera-Medaglia, J., Perron-Welch, F. and Rukundo, O. (2012) *Overview of National and Regional Measures on Access to Genetic Resources and Benefit Sharing. Challenges and Opportunities in Implementing the Nagoya Protocol.* Centre for International and Sustainable Development Law (CISDL). 2nd edn, July. Available at http://cisdl.org/biodiversity-biosafety/public/CISDL_Overview_of_ABS_Measures_2nd_Ed.pdf

Cameron, J. and Brawley, K. (2010) "ABT-594." Department of Chemistry, University of Aberdeen. Available at www.chm.bris.ac.uk/motm/abt/abt.html

Carrizosa, S., Brush, S., Wright, B.D. and McGuire, P.E. (eds.) (2004) *Accessing Biodiversity and Sharing Benefits: Lessons from Implementing the Convention on Biological Diversity.* IUCN Environmental Law Programme. IUCN Environmental Policy and Law Paper No. 54, Gland, Switzerland and Cambridge, UK.

CBD News Special Edition (2002) "The Convention on Biological Diversity from Conception to Implementation." Available at www.cbd.int/doc/publications/CBD-10th-anniversary.pdf

CBD Secretariat (2013) Synthesis of the Online Discussions on Article 10 of the Nagoya Protocol on Access and Benefit-Sharing. UNEP/CBD/ABSEM-A10/1/2. Available at www.cbd.int/doc/?meeting=ABSEM-A10-01

Centeno, J.C. (2009) *La Biopiratería en Venezuela*. Prensa Libre. p. 4. Available at www.rebelion.org/noticias/2009/5/85426.pdf

Chandler, M. (1993) "Biodiversity Convention: Selected Issues of Interest to the International Lawyer." *Colorado Journal of International Environmental Law and Policy* 4(1): 141–175.

Chatterjee, P. (2012) "New Head of CBD: IPR Still Key to Nagoya Protocol on Access and Benefit Sharing." *Intellectual Property Watch Bulletin*. Available at www.ip-watch.org/2012/07/10/cbd-head-ipr-still-key-to-nagoya-protocol-on-access-and-benefit-sharing/

Chen, J.M. (2006) "There's No Such Thing as Biopiracy… And It's a Good Thing Too." *McGeorge Law Review* 37; *Minnesota Public Law Research Paper* No. 05-29. Available at SSRN: http://ssrn.com/abstract=781824

Chiarolla, C., Lapeyre R. and Pirard, R. (2013) *Bioprospecting Under the Nagoya Protocol: A Conservation Booster?* IDDRI. Policy Brief, No. 14/13 November.

Chouchena-Rojas, M., Ruiz, M., Vivas, D. and Winkler, S. (eds.) (2005) *Disclosure Requirements: Ensuring Mutual Supportiveness between the WTO TRIPS Agreement and the CBD*. Gland and Cambridge: IUCN; Geneva: ICTSD. Available at www.ciel.org/Publications/DisclosureRequirements_Nov2005.pdf

CNI (2014) *Study on the Impacts of the Adoption and Implementation of the Nagoya Protocol for Brazilian Industry*. May 2014. pp. 52–60.

Cock M.J.W., van Lenteren, J., Brodeur, J., Barrat, B.I.P., Bigler, F., Bolckmans, K., Consoli, F.N., Haas, F., Mason, P.G. and Parra, J.R.P. (2010) "Do New Access and Benefit Sharing Procedures Under the Convention on Biological Diversity Threaten the Future of Biological Control?" *BioControl* 55(2): 199–218.

Correa, C. (2013) *ITPGRFA: Options to Promote the Wider Application of Article 6.11 of the SMTA and to Enhance Benefit-Sharing*. Legal Opinion. July 2013. The Berne Declaration, The Development Fund. Available at www.evb.ch/fileadmin/files/documents/Biodiversitaet/130731_Juristisches_Gutachten.pdf

Dalton, J. (2013) *Synthetic Biology and the "Omics" Revolution*. The Center for Issue & Crisis Management. United Kingdom. May 2013. Available at www.issue-crisis.com/uploads/Articles/SyntheticBiologyandOmics.pdf

Daly, J.W. (1998) "Thirty Years of Discovering Arthropod Alkaloids in Amphibian Skin." *Journal of Natural Products* 61(1): 162–172.

Daly, J.W. (2003) "Ernest Guenther Award in Chemistry of Natural Products. Amphibian Skin: A Remarkable Source of Biologically Active Arthropod Alkaloids." *Journal of Medicinal Chemistry* 46(4): 445–452.

Daly, J.W., Garraffo, H.M. and Spande, T.F. (2000) "Alkaloids from Frog Skin: The Discovery of Epibatidine and the Potential for Developing Novel Non-opioid Analgesics." *Natural Product Reports* 17(2): 131–135.

Daly, J.W., Spande, T.F. and Garraffo, H.M. (2005) "Alkaloids from Amphibian Skin: A Tabulation of over Eight-hundred Compounds." *Journal of Natural Products* 68(10): 1556–75.

Darst, C.R., Menéndez-Guerrero, P., Coloma, L.A. and Canatella, D.C. (2005) "Evolution of Dietary Specialization and Chemical Defense in Poison Frogs (Dendrobatidae): A Comparative Analysis." *The American Naturalist* 165(1): 56–69.

Darwin, C. (1858) *The Origin of Species: By Means of Natural Selection of the Preservation of Favoured Races in the Struggle for Life*. Signet Classics. 150th Anniversary Edition, September, 2003, pp. 76–127.

David, P. (1985) "Clio and the Economics of QWERTY." *American Economic Review* 75(2): 332–337.

Dawkins, R. (1996) *River Out of Eden. A Darwinian View of Life*. London: Phoenix.

De Jonge, B. (2009) *Plants, Genes and Justice. An Inquiry into Fair and Equitable Benefit Sharing*. Thesis. Wagenigen University, the Netherlands.

De Klemm, C. (1994) "The Problem of Migratory Species in International Law," in Bergesen, H.O., Parmann, G. (eds.) *Green Globe Yearbook of International Co-operation on Environment and Development 1994*, 67–77. Oxford: Oxford University Press.

Dobzhansky, T. (1973) "Nothing in Biology Makes Sense Except in the Light of Evolution." *The American Biology Teacher* 35: 125–129.

Drahos, P. (2014) *Intellectual Property, Indigenous People and their Knowledge*. Cambridge: Cambridge University Press.

Dukat, M. and Glennon, R.A. (2003) "Epibatidine: Impact on Nicotinic Receptor Research." *Cellular and Molecular Neurobiology* 23(3): 365–378.

Echeverri, A. (2010) *Régimen Común de Acceso a los Recursos Genéticos: Biodiversidad y Separación de sus Componentes Intangibles y Tangibles*. Investigación CODI de la Universidad de Antioquia. p. 157. Available at www.leyex.info/magazines/vol67n1496.pdf

Estrella, J., Manoslavas, R., Mariaca, J. and Ribadeneira, M. (2005) *Biodiversidad y Recursos Genéticos. Una Guía para su Uso en el Ecuador*. Ecociencia, DENAREF, Ecuador. Available at www.ecociencia.org/archivos/Biodiversidadyrecursosgeneticos-110922.pdf

ETC Group. CBD COP 12 (2014) *Addressing Synthetic Biology*. Brief. Pyeongchang, Korea.

Febres, M.E. (2002) *La Regulación de Acceso a los Recursos Genéticos en Venezuela*. Centro de Estudios del Desarrollo. Serie Mención Publicación, Caracas, Venezuela.

Felice, F. and Vatiero, M. (2012) "Elinor Ostrom and the Solution to the Tragedy of the Commons." *Il Sussidiario*, June 27, 2012. Available at www.aei.org/article/economics/elinor-ostrom-and-the-solution-to-the-tragedy-of-the-commons/

Fernández Ugalde, J.C. (2007) "Tracking and Monitoring of International Flows of Genetic Resources: Why, How and, Is It Worth the Effort?" in Ruiz, M., Lapeña, I. (eds.) *A Moving Target: Genetic Resources and Options for Tracking and Monitoring their International Flows*, 5–18. ABS Series. Gland: IUCN.

Fitch, R.W., Spande, T.H., Garraffo, M., Herman J.C. and Daly, J.W. (2010) "Phantasmidine: An Epibatidine Congener from the Ecuadorian Poison Frog Epipedobates Anthonyi." *Journal of Natural Products* 73(3): 331–337.

Fore, J., Wiechers, I.R. and Cook-Deegan, R. (2006) "The Effects of Business Practices, Licensing, and Intellectual Property on Development and Dissemination of the Polymerase Chain Reaction: Case Study." *Journal of Biomedical Discovery and Collaboration* 1: 7. doi: 10.1186/1747-5333-1-7

Friedman, T. (2007) *The World Is Flat. A Brief History of the Twenty First Century*. New York: Picador/Farrar, Straus and Giroux.

García, A. and Chirinos, V. (eds.) (1999) "Manual Técnico de Producción de Maca. Recetas Culinarias de la Maca ¡Poderoso Reconstituyente!" *Agronegocios* No. 4: 217–224.

Garrity, G.M., Thomson, D.W., Ussery, N., Paskin, D., Baker, P., Desmeth, D.E., Schindel, D. and Ong, P.S. (2009) "Studies on Monitoring and Tracking Genetic Resources." *Standards in Genomics Sciences*, July 20. Available at www.ncbi.nlm.nih.gov/pmc/articles/PMC3035216/

Genome Canada (2011) *Corporate Plan 2011–12*. Available at www.genomecanada.ca/medias/PDF/EN/CorporatePlan2011-12-english.pdf

Gillis, A.M. (2002) "Serendipity and Sweat in Science. 'Frog Man' Daly Follows Curiosity to Ends of the Earth." *The NIH Record* 44 No. 18. Available at http://nihrecord.od.nih.gov/newsletters/09_03_2002/story01.htm

Glowka, L. (1998) *A Guide to Designing Legal Frameworks to Determine Access to Genetic Resources.* IUCN Environmental Law Centre. Environmental Policy and Law Paper No. 34, Gland, Switzerland and Cambridge, UK.

Glowka, L. (2000) "Bioprospecting, Alien Species and Hydrothermal Vents: Three Emerging Legal Issues in the Conservation and Sustainable Use of Biodiversity." *Tulane Environmental Law Journal* 13: 329–360.

Glowka, L., Burhenne-Guilmin, F. and Synge, H. (1994) *A Guide to the Convention on Biological Diversity.* Gland and Cambridge: IUCN.

GRAIN (2000) *Biodiversity for Sale: Dismantling the Hype about Benefit Sharing. Global Trade and Biodiversity in Conflict – Issue No. 4, April.* Available at www.grain.org/fr/article/entrie s/32-biodiversity-for-sale-dismantling-the-hype-about-benefit-sharing

Grajal, A. (1999) "Régimen de Acceso a los Recursos Genéticos Impone Limitaciones a la Investigación en Biodiversidad en los Países Andinos." *Interciencia* 24(1):63–69.

Greiber, T., Peña Moreno, S., Ahren, M., Nieto Carrasco, J., Chege Kamau, E., Cabrera, J., Olivia, M.J. and Perron-Welch, F. (2013) *An Explanatory Guide to the Nagoya Protocol on Access to Genetic Resources and the Fair and Equitable Sharing of Benefits.* Gland, Switzerland: IUCN.

Halewood, M., Lopez-Noriega, I. and Louafi, S. (2013) *Crop Genetic Resources as a Global Commons. Challenges in International Law and Governance.* Issues in Agricultural Biodiversity. Abingdon and New York: Bioversity International, CGIAR, Earthscan from Routledge.

Hammond, E. (2013) *Biopiracy Watch: A Compilation of Some Recent Cases.* Penang, Malaysia: Third World Network.

Hammond, E. (2014) *Patent Claims on Genetic Resources of Secret Origin. Disclosure Data from Recent International Patent Applications with Related Deposits Under the Budapest Treaty on the International Recognition of the Deposit of Microorganisms for the Purpose of Patent Disclosure.* Third World Network. February, 2014. p. 15. Available at www.twnside.org.sg/title2/ series/bkr/pdf/bkr003.pdf

Heilbroner, R.L. (1979) *The Worldly Philosophers,* 4th edn. New York: Simon and Schuster.

Henne, G. (1997) "'Mutually Agreed Terms' in the Convention on Biological Diversity: Requirements under Public International Law," in Mugabe, J., Barber, C., Henne, G., Glowka, L., La Viña, A. (eds.) *Access to Genetic Resources: Strategies for Benefit Sharing,* 25–53. Kenya: ACTS Press.

Hoagland, E. (1998) "Access to Specimens and Genetic Resources: An Association of Systematics Collections Position Paper." ASCOLL, Washington DC. (ASCOLL is now part of the Natural Science Collections Alliance.)

Hobbelink, H. (1991) *Biotechnology and the Future of World Agriculture.* London: Zed Books.

Hobhouse, H. (1999) *Seeds of Change. Six Plants that Transformed Mankind,* 4th edn. London: Papermac.

IT/ACFS-7 RES/13/Report (2013) *International Treaty on Plan Genetic Resources for Food and Agriculture. Resumed Seventh Meeting of the Ad Hoc Advisory Committee on the Funding Strategy.* April 2013. Available at www.planttreaty.org/sites/default/files/ACFS-7b_Report%20 FINAL.pdf

ITPGRFA (2004) *The International Treaty on Plant Genetic Resources for Food.* Rome: Food and Agriculture Organization of the United Nations. Available at ftp://ftp.fao.org/docrep/ fao/011/i0510e/i0510e.pdf

Jones, W.P., Chin, Y.-W. and Kinghorn, A.D. (2006) "The Role of Pharmacognosy in Modern Medicine and Pharmacy." *Current Drug Targets* 7(3): 247–264.

Kagedan, B.L. (1996) "The Biodiversity Convention, Intellectual Property Rights, and the Ownership of Genetic Resources: International Developments" prepared for the Intellectual Property Policy Directorate Industry Canada. Available at http://iatp.org/files/Biodiversity_Convention_Intellectual_Property_.pdf

Kamau, E.C. and Winter, G. (eds.) (2009) *Genetic Resources, Traditional Knowledge and the Law: Solutions for Access and Benefit Sharing*. London: Earthscan.

Kamau, E.C. and Winter, G. (2013a) "An Introduction to the International ABS Regime and a Comment on its Transposition by the EU." *Law, Environment and Development Journal* 9(2): 108–126. Available at http://ssrn.com/abstract=2387876

Kamau, E.C. and Winter, G. (eds.) (2013b) *Common Pools of Genetic Resources. Equity and Innovation in International Biodiversity Law*. London and New York: Earthscan from Routledge.

Kamau, E.C., Fedder, B. and Winter, G. (2010) "The Nagoya Protocol on Access to Genetic Resources and Benefit Sharing: What is New and What Are the Implications for Provider and User Countries and the Scientific Community?" *Law, Environment and Development Journal* 6(3): 51–65. Available at www.lead-journal.org/content/10246.pdf

Kloppenburg, J. (1988) *First the Seed: The Political Economy of Plant Biotechnology*. Cambridge: Cambridge University Press.

Kloppenburg, J. (2005) *First the Seed: The Political Economy of Plant Biotechnology*, 2nd edn. Science and Technology in Society Series. University of Wisconsin.

Kloppenburg, J. and Kleinman, D.L. (1988) "Seeds of Controversy: National Property versus Common Heritage," in Kloppenburg, J. (ed.) *Seeds and Sovereignty: The Use and Control of Plant Genetic Resources*, 173–203. Chapel Hill, NC: Duke University Press.

Krauss, L. (2013) "Deafness at Doomsday." *New York Times* (The Opinion Pages), January 15. Available at www.nytimes.com/2013/01/16/opinion/deafness-at-doomsday.html?_r=0

Krugman, P. (2014a) "Why Economics Failed." *New York Times* (The Opinion Pages), May 1. Available at nyti.ms/1kz4iZ7

Krugman, P. (2014b) "Point of No Return." *New York Times* (The Opinion Pages), May 15. Available at www.nytimes.com/2014/05/16/opinion/krugman-points-of-no-return.html?_r=1

Laird, S. (1993) "Contracts for Biodiversity Bioprospecting," in Reid, W., Laird, S., Meyer, C., Gámez, R., Sittenfeld, A., Janzen, D., Gollin, M., Juma, C. (eds.) *Biodiversity Prospecting. Using Genetic Resources for Sustainable Development*. Washington DC: World Resources Institute.

León, C. (1986) "Un proyecto en marcha." *AgroNoticias*, September 1986, 22–23.

Lewis, W., Lamas, G., Vaisberg, A., Corley, D.G. and Sarasara, C. (1999) "Peruvian Medicinal Plant Sources of New Pharmaceuticals (ICBG – Peru)," in Rosenthal, J. (ed.) "Drug Discovery, Economic Development and Conservation: The International Cooperative Biodiversity Groups". *Pharmaceutical Biology* 37, Supplement, Swets & Zeitlinger, the Netherlands.

Li, G., Ammermann, U. and Quirós, C. (2001) "Glucosinolate contents in Maca (*Lepidium peruvianum* Chacón) Seeds, Sprouts, Mature Plants and Several Derived Commercial Products." *Economic Botany* 55(2): 255–262.

Lohan, D. and Johnston, S. (2005) *Bioprospecting in Antarctica*. Yokohama, Japan: UNU-IAS.

Lukács, B. *A Note to the Lost Books of Aristotle*. Available at www.rmki.kfki.hu/~lukacs/ARISTO3.htm

Luo, Y., Lee, J.K. and Zhao, H. (2012) "Challenges and Opportunities in Synthetic Biology for Chemical Engineers." *Chemical Engineering Science* 103: 115–119. Available at http://dx.doi.org/10.1016/j.ces.2012.06.013

Manheim, B. (2014) "Nagoya Protocol Spurs New and More Stringent Requirements for Prior Informed Consent and Benefit Sharing for Research and Commercial Activities Involving Genetic Resources from Plants, Animals and Microorganisms." October 17, 2014. Available at www.mondaq.com/unitedstates/x/347698/Life+Sciences+Biotechnology/Restrictions+Governing+International+Trade+in+Genetic+Resources+Enter+Into+Force

Mansur, A. and Cavalcanti, K. (1999) "Xenofobia na Selva: Paranoia Envolvendo Biopirataria Prejudica Pesquisas Cientificas com Especies Brasileiras." *Revista Veja*, Ed. 1611, Año 32, No. 32–33, Agosto 1999, pp. 114–118.

Markandya, A. and Nunes, P. (2012) *Is the Value of Bioprospecting Contracts Too Low?* Nota di Lavoro. Fondazione Eni Enrico Mattei. Available at www.feem.it/userfiles/attac h/20101213150154NDL2010-154.pdf

Marrero-Girona, G. and Vogel, J.H. (2012) "Can 'Monkey Business' Resolve the Most Contentious Issue in the Convention on Biological Diversity?" *International Journal of Psychological Studies* 4(1): 55–65. Available at www.ccsenet.org/journal/index.php/ijps/article/view/15456/10582

Marshall, A. (1890) *Principles of Economics.* London: Macmillan & Co., Ltd. Available at www.econlib.org/library/Marshall/marP.html

Martinez-Alier, J. (2002) *The Environmentalism of the Poor: A Study of Ecological Conflicts and Valuation.* Cheltenham and Northampton: Edward Elgar Publishing.

Martinez-Alier, J. (2005) *El Ecologismo de los Pobres. Conflictos Ambientales y Lenguajes de Valoración*, 3rd edn. Madrid: Icaria Ed.

May, C. (2010) *The Global Political Economy of Intellectual Property Rights. The New Enclosures*, 2nd edn. London: Routledge.

Mazoomdaar, J. (2014) "Centre Sits on Royalty Slabs for Bio Resources, Loses Rs 25,000 cr a Year" *NATION*, November 19. Available at http://indianexpress.com/article/india/india-others/centre-sits-on-royalty-slabs-for-bio-resources-loses-rs-25000-cr-a-year/

McGraw, D.M. (2000) "The Story of the Biodiversity Convention: Origins, Characteristics and Implications for Implementation," in Le Prestre, P.G (ed.) *The Convention on Biological Diversity and the Construction of a New Biological Order*, 9–43. Aldershot, UK: Ashgate.

McManis, C.R. (2004) "Fitting Traditional Knowledge Protection and Biopiracy Claims into the Existing Intellectual Property and Unfair Competition Framework," in Burton, O. (ed.) *Intellectual Property and Biological Resources*, 425–510. London: Marshall Cavendish International.

McManis, C.R. (2007) (ed.) *Biodiversity and the Law. Intellectual Property, Biotechnology and Traditional Knowledge.* London and Sterling, VA: Earthscan.

Memorandum from Joshua Sarnoff to Public Interest Intellectual Property Advisors (PIIPA) (2004) "Compatibility with Existing International Intellectual Property Agreements of Requirements for Patent Applications to Disclose Origins of Genetic Resources and Traditional Knowledge and Evidence of Legal Access and Benefit Sharing." Available at www.piipa.org/index.php?option=com_content&view=article&id=91

Mindreau, M. (2005) *Del GATT a la OMC (1947–2005): La Economía Política Internacional del Sistema Multilateral de Comercio.* Universidad del Pacífico. Lima, Peru.

Mooney, P.R. (1979) *Seeds of the Earth: A Public of Private Resources?* Ottawa: Canadian Council for International Cooperation and the International Coalition for Development Action.

Mooney, P.R. (1983) *The Law of the Seed – Another Development and Plant Genetic Resources.* Dag Hammarskjöld Foundation. Available at www.dhf.uu.se/pdffiler/83_1-2.pdf

Moore v. Regents of the University of California (1990) 51 C3d 120 favors its status as res nullius. Available at http://online.ceb.com/CalCases/C3/51C3d120.htm

Myers, C.W., Daly, J.W. and Malkin, B. (1978). "A Dangerously Toxic New Frog (Phyllobates) Used by Embera Indians of Western Colombia, with Discussion of Blowgun Fabrication and Dart Poisoning." *Bulletin of the American Museum of Natural History* 161(2): 307–366.

Natural Justice, The Berne Declaration (2013) *Access or Utilisation – What Triggers User Obligations? A Comment on the Draft Proposal of the European Commission on the Implementation of the Nagoya Protocol on Access and Benefit Sharing.* Available at http://naturaljustice.org/wp-content/uploads/pdf/Submission-EU-ABS-Regulation.pdf

Newman, D.J. and Cragg, G.M. (2013) "Natural Products as Sources of New Drugs over the 30 Years from 1981 to 2010." NIH. Public Access. Authors' Manuscript. July 24, 2013. Available at www.ncbi.nlm.nih.gov/pmc/articles/PMC3721181/pdf/nihms356104.pdf

Nirogi, R., Goura, V. and Abraham, R. (2013) "$\alpha4\beta2\star$ Neuronal Nicotinic Receptor Ligands (Agonist, Partial Agonist and Positive Allosteric Modulators) as Therapeutic Prospects for Pain." *European Journal of Pharmacology* 712(1–3): 22–29.

Nnadozie, K., Lettington, R., Bruch, C., Bass, S. and King, S. (2003) *African Perspectives on Genetic Resources. A Handbook on Laws, Policies and Institutions.* African Union, ELI, SEAPRI. Kenya.

Ochoa, C. and Ugent, D. (2001) "Maca (*Lepidium meyenii* Walp.: Brassicaceae): A Nutritious Root Crop of the Central Andes." *Economic Botany* 55(3): 344–345.

Oduardo-Sierra, O., Hocking, B.A. and Vogel, J. H. (2012) "Monitoring and Tracking the Economics of Information in the Convention on Biological Diversity: Studied Ignorance (2002–2011)." *Journal of Politics and Law* 5(2): 29–39.

Oldham, P. (2004) "Global Status and Trends in Intellectual Property Claims: Genomics, Proteomics and Biotechnology." Submission to the Executive Secretary of the Convention on Biological Diversity. Center for Economic and Social Aspects of Genomics. United Kingdom. Available at www.cesagen.lancs.ac.uk/resources/docs/genomics-final.doc

Oldham, P. (2009) "An Access and Benefit Sharing Commons? The Role of Common/Open Source Licenses in the International Regime on Access to Genetic Resources and Benefit Sharing." Research Documents. Initiative for the Prevention of Biopiracy. Year IV, No. 12, July 2009. Available at www.biopirateria.org/download/documentos/investigacion/rrggs/serie_iniciativa12.pdf

Oldham, P., Hall, S. and Forero, O. (2013) "Biological Diversity in the Patent System." *PLoS ONE* 8(11):6.

Pastor, S. and Ruiz, M. (2008) *The Development of an International Regime on Access to Genetic Resources and Fair and Equitable Benefit Sharing in a Context of New Technological Developments.* Initiative for the Prevention of Biopiracy. Year III, No. 9. Available at www.cbd.int/abs/doc/serie-iniciativa-2009-04-en.pdf

Pauli, G.F., Chen, S.N., Brent Friesen, J., McAlpine, J.M. and Birgit, U.J. (2012) "Analysis and Purification of Bioactive Natural Products: The AnaPurNa Study." *Journal of Natural Products* 75(6): 1243–1255.

Perrault, A. and Oliva, M.J. (2005) *Dialogue on Disclosure Requirements: Incorporating the CBD Principles in the TRIPS Agreement on the Road to Hong Kong WTO Public Symposium,* ICTSD/CIEL/IDDRI/IUCN/QUNO, Geneva, April 21, 2005. Available at www.ictsd.org/downloads/2008/12/meeting-report.pdf

Pistorious, R. (1997) *Scientists, Plants and Politics. A History of the Plant Genetic Resources Movement.* Rome: International Plant Genetic Resources Institute.

Ploetz, C. (2005) "ProBenefit: Process-oriented Development for a Fair Benefit-sharing Model for the Use of Biological Resources in the Amazon Lowland of Ecuador," in Feit, U., Von den Driesch, M., Lobin, W. (eds.) *Access and Benefit-Sharing of Genetic Resources Ways and Means for Facilitating Biodiversity Research and Conservation while Safeguarding ABS Provisions*, 97–101. Report of an International Workshop in Bonn, Germany, convened by the German Federal Agency for Nature Conservation. November 8–10, 2005. Available at www.bfn.de/fileadmin/MDB/documents/service/skript163.pdf

PL 7735/14. "Projeto da biodiversidade vai a comissão geral com várias polêmicas em aberto," (November 6, 2014). Available at www2.camara.leg.br/camaranoticias/noticias/POLITICA/477144-PROJETO-DA-BIODIVERSIDADE-VAI-A-COMISSAO-GERAL-COM-VARIAS-POLEMICAS-EM-ABERTO.html

Pulgar, J. (1978) "La Maca y el uso de la región Puna VIII." *Expreso*, July 1978.

Quezada, F. (2007) *Status and Potential of Bioprospecting Activities in Latin America and the Caribbean.* Serie Medio Ambiente y Desarrollo. CEPAL. No. 132, Santiago de Chile.

Rausser, G.C. and Small, A.A. (2000) "Valuing Research Leads: Bioprospecting and the Conservation of Genetic Resources." *Journal of Political Economy* 108(1):173–206.

Renner, S.C., Neumann, D., Burkart, M., Feit, U., Giere, P., Groger, A., Paulsch, A., Paulsch, C., Sterz, M. and Vohland, K. (2012) "Import and Export of Biological Samples from Tropical Countries – Considerations and Guidelines for Research Teams." *Organisms Diversity & Evolution* 12(1):81–98.

Ribadeneira, M. (2008) "La Biopiratería, El Desafío de Construir un Camino entre una Acusación Política y una Categoría Legal," in Lottici, M.V. (ed.) *Conservación de la Biodiversidad y Política Ambiental. Sexta Convocatoria, Premio de Monografía Adriana Schiffrin 2007, Trabajos Premiados*, 87–117. Buenos Aires: Fundación Ambiente y Recursos Naturales.

Robinson, D.F. (2010) *Confronting Biopiracy: Challenges, Cases and International Debates.* London and Washington DC: Earthscan.

Rosell, M. (1997) "Access to Genetic Resources: A Critical Approach to Decision 391 'Common Regime on Access to Genetic Resources'." *RECIEL* 3(3): 274–283.

Rosenthal, J. (ed.) (1999) "Drug Discovery, Economic Development and Conservation: The International Cooperative Biodiversity Groups." *Pharmaceutical Biology* 37 (Supplement), Swets & Zeitlinger, the Netherlands.

Rosenthal, J., Beck, D.A., Bhat, A., Biswas, J., Brady, L., Bridbord, K., Collins, S., Cragg, G., Edwards, J., Fairfield, A., Gottlieb, M. and Gschwind, L.A. (1999) "Combining High Risk Science with Ambitious Social and Economic Goals". *Pharmaceutical Biology* 37 (Supplement), Swets & Zeitlinger, the Netherlands. pp. 6–21.

Ruiz, M. (1999) *Acceso a Recursos Genéticos. Propuestas e Instrumentos Jurídicos.* Sociedad Peruana de Derecho Ambiental, Lima, Peru, pp. 7–35.

Ruiz, M. (2003) *¿Es Necesario un Nuevo Marco Jurídico para la Bioprospección en la región Andina: Breve Reflexión Crítica de la Decisión 391?* Serie de Política y Derecho Ambiental. No. 14, Lima, Peru.

Ruiz, M. (2008) *Guía Explicativa de la Decisión 391 y una Propuesta Alternativa para Regular el Acceso a los Recursos Genéticos en la Región Andina.* GTZ, SPDA, The MacArthur Foundation, Lima, Peru.

Ruiz, M. (2011) *Diseño de un Plan de Fortalecimiento de Capacidades Institucionales en el Tema de Acceso a los Recursos Genéticos Asociados a los Conocimientos Tradicionales.* Diagnóstico Regional y Anexos. Documento de trabajo. BIOCAN. Comunidad Andina. December

22, 2011. Available at http://biocan.comunidadandina.org/biocan/images/documentos/TallerARG/diagnostico_abs_documento_trabajo.doc

Ruiz, M. and Lapeña, I. (2007) (eds.) *A Moving Target: Genetic Resources and Options for Tracking and Monitoring International Flows.* ABS Series. IUCN Environmental Policy and Law Paper No. 67/3, Gland, Switzerland.

Ruiz, M., Vogel, J. and Zamudio, T. (2010) *Logic Should Prevail: A New Theoretical and Operational Framework for the International Regime on Access to Genetic Resources and the Fair and Equitable Sharing of Benefits.* Initiative for the Prevention of Biopiracy. SPDA. Research Documents. Year V, No. 13, 2010, Lima, Peru. Available at www.planttreaty.org/sites/default/files/logic_ABS_biopiracy.pdf

Saavedra, L.A. (1999) "Invasion of the Frog-snatchers." *New Internationalist* 311. Available at http://newint.org/columns/currents/1999/04/01/ecuador/

Sampford, C.J.G. (2006) *Retrospectivity and the Rule of Law.* Oxford: Oxford University Press.

Samuelson, P. and Nordhaus, W. (2005) *Economics*, 18th edn. New York: McGraw-Hill.

Saporito, R.A., Donnelly, M.A., Spande, T.F. and Garraffo, H.M. (2012) "A Review of Chemical Ecology in Poison Frogs." *Chemoecology* 22(3): 159–168.

Schei, J. and Tvedt, M.W. (2010) *Genetic Resources in the CBD: The Wording, the Past, the Present and the Future.* FNI report. Available as UNEP/CBD/WG-ABS/9/INF/1 at www.cbd.int/doc/meetings/abs/abswg-09/information/abswg-09-inf-01-en.pdf

Schrodinger, E. (1944) *What is Life?* Manuscript available at http://whatislife.stanford.edu/LoCo_files/What-is-Life.pdf

Sedjo, R. (1988) "Property Rights and the Protection of Plant Genetic Resources," in Kloppenburg, J.R. (ed.) *Seeds and Sovereignty*, 293–314. Durham, NC: Duke University Press.

Sedjo, R. (1989) "Property Rights for Plants." *Resources* (Fall, 97): 1–4.

Simpson, R.D., Sedjo, R.A. and Reid, W.J. (1996) "Valuing Biodiversity for Use in Pharmaceutical Research." *Journal of Political Economy* 104(1): 163–185.

Smith, A. (2007) *The Wealth of Nations* (1776). Available at http://metalibri.wikidot.com/title:an-inquiry-into-the-nature-and-causes-of-the-wealth-of

Spande, T.F., Garraffo, H.M., Edwards, M.W., Yeh, H.J.C., Pannell, L. and Daly, J.W. (1992) "Epibatidine: A Novel (Chloropyridyl) Azabicycloheptane with Potent Analgesic Activity from an Ecuadoran Poison Frog." *Journal of the American Chemical Society* 114(9): 3475–3478.

SPDA/CDA-UICN (1994) Hacia un Marco Legal para Regular el Acceso a los Recursos Genéticos en el Pacto Andino: Posibles Elementos para una Decisión del Pacto Andino sobre Acceso a los Recursos Genéticos. Reporte Técnico Legal preparado por el Centro de Derecho Ambiental de la Unión Mundial para la Naturaleza (UICN) para la Junta del Acuerdo de Cartagena con la asistencia técnica de la Sociedad Peruana de Derecho Ambiental (SPDA). JUN/REG.ARG/I/dt.4, 31 Octubre de 1994, in Ruiz, M. (1999) *Acceso a Recursos Genéticos. Propuestas e Instrumentos Jurídicos.* Sociedad Peruana de Derecho Ambiental, Lima, Peru, pp. 7–35.

Stenton, G. (2003) "Biopiracy within the Pharmaceutical Industry: A Stark Illustration of Just How Abusive, Manipulative and Perverse the Patenting Process can be Towards Countries of the South." *Hertfordshire Law Journal* 1(2): 30–40.

Stoll P.T. (2013) "ABS, Justice, Pools and the Nagoya Protocol," in Chege Kamau, E., Winter, G. (eds.) *Common Pools of Genetic Resources. Equity and Innovation in International Biodiversity Law*, 305–314. New York: Routledge.

Stone, C. (1972) "Should Trees have Standing? Towards Legal Rights for Natural Objects." *Southern California Law Review* No. 45, p. 450.

Stone, C.D. (1995) "What to Do about Biodiversity, Property Rights, Public Goods and the Earth's biological Riches." *Southern California Law Review* No. 68, pp. 577–605.

Sukhwani, A. (1995) *Patentes Naturistas*. Madrid: Oficina Española de Patentes.

Suneetha, M.S. and Pisupati, B. (2009) *Benefit Sharing in ABS: Options and Elaborations*. Nairobi: UNEP, UNU-IAS.

Swanson, T.M. (1992) "The Economics of the Biodiversity Convention." Norwich: CSERGE, School of Environmental Sciences, University of East Anglia.

Swanson, T. (1997) *Global Action for Biodiversity*. Abingdon, UK: Earthscan.

Swanson, T.M., Pearce, D.W. and Cervigni, R. (1994) *The Appropriation of the Benefits of Plant Genetic Resources for Agriculture: An Economic Analysis of the Alternative Mechanism for Biodiversity Conservation*. Rome: Secretariat of the FAO Commission on Plant Genetic Resources.

Takushi, S. (2013) "Biological Prospectors, Pirates, Pioneers, and Punks in the Andes Mountains: An Examination of Scientific Practice in the Andean Community of Nations." *Honors Projects*, (2013). Paper 16. Available at http://digitalcommons.iwu.edu/intstu_honproj/16

ten Brink, P. (2009) "Rewarding Benefits through Payments and Markets," in *The Economics of Ecosystems and Biodiversity*. TEEB.

ten Kate, K. and Laird, S. (eds.) (1999) *The Commercial Use of Biodiversity: Access to Genetic Resources and Benefit Sharing*. London, UK and Sterling, VA: Earthscan.

ten Kate, K., Touche, L. and Collins, A. (1998) *Benefit Sharing Case Study: Yellowstone Park and Diversa Corporation*. Submission to the Executive Secretary of the Convention on Biological Diversity by the Royal Botanic Gardens, Kew, April 22, 1998. Available at www.cbd.int/financial/bensharing/unitedstates-yellowstonediversa.pdf

Thampi, S. *Bioinformatics* (date not indicated). Available at http://arxiv.org/ftp/arxiv/papers/0911/0911.4230.pdf

The International Civil Society Working Group on Synthetic Biology (2011) *A Submission to the Convention on Biological Diversity's Subsidiary Body on Scientific, Technical and Technological Advice (SBSTTA) on the Potential Impacts of Synthetic Biology on the Conservation of Biodiversity*. October 17, 2011. Available at www.cbd.int/doc/emerging-issues/Int-Civil-Soc-WG-Synthetic-Biology-2011-013-en.pdf

Third World Network (2013) *Costa Rica's INBIO Nearing Collapse, Surrenders its Biodiversity Collections and Seeks Government Bailout*. April 2013. Available at www.twnside.org.sg/title2/biotk/2013/biotk130401.htm

Tidwell, J. (2002) "Raiders of the Forest Cures." *Zoogoer* (Sept/Oct): 14–21. Available at http://static.squarespace.com/static/5244b0aee4b045a38d48f8b0/t/5339967ce4b041f3867ab786/1396283004015/Raiders%20of%20the%20Forest%20Cures.pdf

Tobin, B. (1997) "Certificates of Origin: A Role for IPR Regimes in Securing Prior Informed Consent," in Mugabe, J., Barber, C., Henne, G., Glowka L., La Viña, A. (eds.) *Access to Genetic Resources: Strategies for Benefit Sharing*. Kenya: ACTS Press. Available at www.academia.edu/6636676/Certificates_of_Origin_A_role_for_IPR_Regimes_in_Securing_Prior_Informed_Consent

Tobin, B. (2009) "Setting Protection of TK to Rights – Placing Human Rights and Customary Law at the Heart of TK Governance," in Kamau, E.C., Winter, G. (eds.) *Genetic Resources, Traditional Knowledge and the Law. Solutions for Access and Benefit Sharing*, 101–118. London and Sterling, VA: Earthscan.

Tobin, B. and Taylor, E. (2009) *Across the Great Divide: A Case Study of Complementarity and Conflict between Customary Law and TK Protection Legislation in Peru*. Research Documents.

Initiative for the Prevention of Biopiracy. SPDA. Year IV No. 11. May 2009. Available at www.biopirateria.org/documentos/Serie%20Iniciativa%2011.pdf

Tvedt, M.W. (2011) *A Report from the First Reflection Meeting on the Global Multilateral Benefit Sharing Mechanism.* FNI Report 10/2011. Report Lysaker, Norway: Fridtjof Nansen Institute, ABS Capacity Development Initiative – GIZ. Available at www.fni.no/doc&pdf/ FNI-R1011.pdf

Tvedt, M.W. and Fauchald, O.K. (2011) "Implementing the Nagoya Protocol on ABS: A Hypothetical Case Study on Enforcing Benefit Sharing in Norway." *The Journal of World Intellectual Property* 14 (5): 383–402,

UNEP/CBD/ICNP/3/INF/4 Synthesis of the Online Discussion on Article 10 of the Nagoya Protocol on Access and Benefit-sharing, April 2014. Available at www.cbd.int/ doc/?meeting=ABSEM-A10-01

UNEP/CBD/WG-ABS/9/INF/15, March 10, 2010, *Proceedings of the Seminar "Barcoding of Life: Society and Technology Dynamics – Global and National Perspectives".* Submitted by the International Development Research Centre of Canada. Available at www.cbd.int/doc/ meetings/abs/abswg-09/information/abswg-09-inf-15-en.pdf

United Nations. *Nagoya Protocol on Access to Genetic Resources and the Fair and Equitable Sharing of Benefits Arising from their Utilization to the Convention on Biological Diversity.* Opened for signature on October 29, 2010 (entered into force on October 12, 2014). Available at www.cbd.int/abs/text/default.shtml

United Nations. *Convention on Biological Diversity.* Opened for signature on June 5, 1992 (entered into force on December 29, 1993). Available at www.cbd.int/convention/ text/

Vogel, J.H. (1990) "Intellectual Property and Information Markets: Preliminaries to a New Conservation Policy." *CIRCIT Newsletter* (May): 6.

Vogel, J.H. (1991) "The Intellectual Property of Natural and Artificial Information." *CIRCIT Newsletter* (June): 7.

Vogel, J.H. (1992) *Privatisation as a Conservation Policy.* Melbourne, Australia: Centre for International Research on Communication and Information Technologies.

Vogel, J.H. (1994) *Genes for Sale.* New York: Oxford University Press.

Vogel, J.H. (1997) "White Paper: The Successful Use of Economic Instruments to Foster the Sustainable Use of Biodiversity: Six Cases from Latin America and the Caribbean." *Biopolicy Journal* 2 (Paper 5). Available at www.bioline.org.br/request?py97005

Vogel, J. (ed.) (2000) *The Biodiversity Cartel: Transforming Traditional Knowledge into Trade Secrets.* CARE. Proyecto SUBIR, Quito, Ecuador. Available at http://josephhenryvogel. com/cartelengl_files/01.pdf

Vogel, J.H. (2005) "Sovereignty as a Trojan Horse: How the Convention on Biological Diversity Morphs Biopiracy into Biofraud," in Hocking, B.A. (ed.) *Unfinished Constitutional Business? Rethinking Indigenous Self-Determination,* 228–247. Australia: Aboriginal Studies Press.

Vogel, J.H. (2007a) "Reflecting Financial and Other Incentives of the TMOIFGR: The Biodiversity Cartel," in Ruiz, M., Lapeña, I. *A Moving Target: Genetic Resources and Options for Tracking and Monitoring their International Flows,* 47–74. Gland, Switzerland: IUCN. English: http://data.iucn.org/dbtw-wpd/edocs/EPLP-067-3.pdf. Spanish: http:// cmsdata.iucn.org/downloads/eplp_67_3_sp.pdf. French: http://cmsdata.iucn.org/ downloads/eplp_67_3_fr.pdf

Vogel, J.H. (2007b) "From the Tragedy of the Commons to the Tragedy of the Common Place: Analysis and Synthesis through the Lens of Economic Theory," in McManis, C.

(ed.) *Biodiversity and the Law. Intellectual Property, Biotechnology and Traditional Knowledge*, 92–115. London and Sterling, VA: Earthscan,

Vogel, J.H. (2008) "Nothing in Bioprospecting Makes Sense Except in the Light of Economics," in Sunderland, N., Graham, P., Isaacs, P., McKenna, B. (eds.) *Toward Humane Technologies: Biotechnology, New Media and Ethics*, 65–74. Rotterdam: Sense Publishers Series.

Vogel, J.H. (2012) "Architecture by Committee and the Conceptual Integrity of the Nagoya Protocol," in Ruiz, M., Vernooy, R. (eds.) *The Custodians of Biodiversity. Sharing Access to and Benefits of Genetic Resources*, 184. New York: Earthscan.

Vogel, J.H. (2013) "The Tragedy of Unpersuasive Power: The Convention on Biological Diversity as Exemplary." *International Journal of Biology* 5(4): 44–54. Available at www.ccsenet.org/journal/index.php/ijb/article/view/30097/18019.

Vogel, J.H., Robles, J., Gomides, C. and Muñiz, C. (2008) "Geopiracy as an Emerging Issue in Intellectual Property Rights: The Rationale for Leadership by Small States." *Tulane Environmental Law Journal* 21(Summer): 391–406.

Vogel, J.H., Alvarez, N., Quiñonez, N., Medina, J., Perez, D., Arocho, A., Val-Verniz, N., Fuentes, R., Marrero-Girona, G., Valcárcel, E. and Santiago, J. (2011) "The Economics of Information, Studiously Ignored in the Nagoya Protocol, on Access to Genetic Resources and Benefit Sharing." *Law, Environment and Development Journal* 7(1): 52. Available at www.lead-journal.org/content/11052.pdf

Walsh, V. and Goodman, J. (1999) "Cancer Chemotherapy, Biodiversity, Public and Private Property: The Case of the Anti-Cancer Drug Taxol." *Social Science & Medicine* 49: 1251–1255.

Watanabe, K.N. and Teh, G.H. (2011) "Wanted: Bioprospecting Consultants." *Nature Biotechnology* 29(10): 873–875.

Watson J.D. and Crick, F.H.C. (1953) "Molecular Structure of Nucleic Acids: A Structure for Deoxyribose Nucleic Acid." *Nature* 171: 737–738.

West, S. (2012) "Institutionalised Exclusion: The Political Economy of Benefit Sharing and Intellectual Property." *Law, Environment and Development Journal* 8(1): 19. Available at www.lead-journal.org/content/12019.pdf

Williams, M., Garraffo, H.M. and Spande, T.F. (2009) "Epibatidine: From Frog Alkaloid to Analgesic Clinical Candidates. A Testimonial to 'True Grit'!" *Heterocycles* 79(1): 207–217.

Wilson, E.O. (1984) *Biophilia*. Cambridge, MA: Harvard University Press.

Wilson, E.O. (1992) *The Diversity of Life*. New York: W.W. Norton & Company.

Wilson, E.O. (2002) *The Future of Life*. New York: Random House.

Wilson, E.O. (2012) *The Social Conquest of Earth*. New York: W.W. Norton.

Wilson, E.O. (2014) *The Meaning of Life*. New York: W.W. Norton.

Winands, S. and Holm-Muller, K. (2014) "Eco-regional Cartels on the Genetic Resource Market and the Case of the Andean Community's Legislation." Institute for Food and Resources Economics. Agricultural and Resources Economics, Discussion Paper 2014: 2.

Winter, G. (2009) "Towards Regional Common Pools of GRs – Improving Equity and Fairness in ABS," in Kamau, C.E., Winter, G. (eds.) *Genetic Resources, Traditional Knowledge and the Law. Solutions for Access and Benefit Sharing*. London: Earthscan.

Winter, G. (2013) "Knowledge Commons, Intellectual Property and the ABS Regime," in Chege Kamau, E., Winter, G. (eds.) *Common Pools of Genetic Resources. Equity and Innovation in International Biodiversity Law*. London and New York: Routledge.

WIPO (2003) "Delegation of Peru. Patents referring to *Lepidium meyenii* (Maca): Responses of Peru." Intergovernmental Committee on Intellectual Property and Genetic Resources,

Traditional Knowledge and Folklore of the World Intellectual Property Organization. Geneva, July 2003. Available at www.wipo.int/edocs/mdocs/tk/en/wipo_grtkf_ic_5/wipo_grtkf_ic_5_13.doc

WIPO (2013) *Draft Intellectual Property Guidelines for Access to Genetic Resources and Equitable Sharing of the Benefits Arising from their Utilization.* Consultation Draft, February, 2013, p. 21. Available at www.cbd.int/financial/mainstream/wipo-guidelines.pdf

WIPO (2014) *Traditional Cultural Expressions.* Available at www.wipo.int/tk/en/folklore/

Wong, T. and Dutfield, G. (eds.) (2011) *Intellectual Property and Human Development. Concerns, Trends and Future Scenarios.* Cambridge: Public Interest Intellectual Property Advisors.

Zapata, B. (2004) *La Experiencia Boliviana en la Aplicación de la Decisión 391: Régimen Común sobre Acceso a los Recursos Genéticos.* Ministerio de Desarrollo Sostenible. La Paz, Bolivia.

Annotated filmography

An asteroid as the causation of the last mass extinction is accepted science. An analogy with the asteroid lies in (a) the THIPPO of mass extinction today and (b) the amplifications effects of a seemingly small difference in trajectories. See bajavose, KT Asteroid & Dinosaurs Extinction, 2013. www.youtube.com/watch?v=ubBebEywNmE

HIPPO is superimposed on the screen as Professor E.O. Wilson explains the agents of mass extinction. The fact that he and other biologists have been explaining those agents for more than half a century legitimizes the derivative THIPPPO, where T is the tragedy of unpersuasive power. See NRDCflix, "E.O. Wilson & Elizabeth Kolbert," 2008. www.youtube.com/watch?v=GIlvstjsp8I

Some events were realized as historic as they were happening. The interviews with the movers and shakers of the Convention on Biological Diversity give us a glimpse into the solemn responsibility they felt as well as their apprehensions. See greentv, "Reflecting on Rio – Looking Back to 1992," 2012. www.youtube.com/watch?v=68WECTt_DfU

The economics of information derives from an asymmetric cost structure of production. The implications are sufficiently sweeping to justify specialization in its application at the level of graduate education. See Maastricht University, "Introduction master Infonomics @ Maastricht University," 2011. www.youtube.com/watch?v=pyn3stjN1bs

Quite literally genetic resources are information. The narrator ends the video advising the audience not to listen to those who cannot say "what is information and what isn't." See Shane Killian, "Evolution CAN Increase Information," 2011. www.youtube.com/watch?v=gJVjTh98aHU

The adherence of the COPs to the wrong definition of genetic resources is fully explicable. However, its mere identification will never be enough to correct it; we must also be aware of the human inclination to defend what is known to be mistaken. See Marco Torres, "Psychology of the 'Sunk Cost Effect,'" 2014. www.youtube.com/watch?v=DDIsR4J1jh4

Index